Child and Adolescent Psychiatry: A Post-COVID Era?

Child and Adolescent Psychiatry: A Post-COVID Era?

Guest Editor

Matteo Alessio Chiappedi

MDPI

Basel • Beijing • Wuhan • Barcelona • Belgrade • Novi Sad • Cluj • Manchester

Guest Editor
Matteo Alessio Chiappedi
Child Neuropsychiatry Unit
ASST Pavia
Pavia
Italy

Editorial Office
MDPI AG
Grosspeteranlage 5
4052 Basel, Switzerland

This is a reprint of the Special Issue, published open access by the journal *Children* (ISSN 2227-9067), freely accessible at: https://www.mdpi.com/journal/children/special_issues/325IP1Q923.

For citation purposes, cite each article independently as indicated on the article page online and as indicated below:

Lastname, A.A.; Lastname, B.B. Article Title. *Journal Name* **Year**, *Volume Number*, Page Range.

ISBN 978-3-7258-3635-2 (Hbk)
ISBN 978-3-7258-3636-9 (PDF)
https://doi.org/10.3390/books978-3-7258-3636-9

Contents

Preface

When I started this Special Issue, the SARS-CoV-2 pandemic was not over yet. When the storm finally ended, it looked like the most prevalent desire was to try to forget those days, characterized by too much pain, mourning, and suffering. A number of scholars, however, accepted my invitation and found the strength and courage to contribute to this Special Issue, showing not only their will to help children and adolescents with psychiatric issue but (and most importantly) that they can also overcome their legitimate fears to achieve what doctors have been doing for as long as this profession has existed: studying ways to help their patients live as best as possible. I thank all of them for gifting me with their examples. This reprint is not merely a collection of papers but wants to be a concrete symbol of hope.

Matteo Alessio Chiappedi
Guest Editor

children

MDPI

Editorial

Child and Adolescent Psychiatry: Unmet Needs Before and After the COVID-19 Pandemic

Matteo Chiappedi

Child and Adolescent Neuropsychiatry Unit, ASST Pavia, 27029 Vigevano, Italy; matteo.chiappedi@unipv.it

The SARS-CoV-2 pandemic formally began on 11 March 2020, when the World Health Organization declared it as an evolution of the Public Health Emergency of International Concern (PHEIC). Only in May 2023 did the World Health Organization reduce the classification to an "established and ongoing health issue". What happened in these 3 years had not been foreseen and provided undeniable evidence that modern day medicine still has limitations, but also holds great potential (with the COVID-19 vaccines being the symbol of the latter).

A number of changes in everyday life were necessary; in many cases, these changes were also established by law. Many of them could be seen as potential negative factors for children and adolescents' psychological wellbeing. One could divide these factors into three main groups—direct consequences of the pandemic (e.g., having parents or friends ill or fearing to be infected, difficulties in receiving medical and psychological support, etc.); changings in protective everyday experiences (e.g., avoiding friends, sports and leisure activities of all kinds and replacing them with solitary activities due to "stay-at-home" orders; school closures with distance learning often not readily and sufficiently applied; etc.); blame posed on "normal" requests (e.g., in Italy, the need to protect fragile and older people led to blame being assigned to children and adolescents who tried to live their lives to the best of their ability within existing limitations).

Among the 17 contributions submitted to this Special Issue, 11 were accepted for publication (64.7%). These papers can be divided into two main groups according to their topic.

The first group investigated more general aspects of child and adolescent psychiatry, which was highly relevant both before and after the SARS-Cov-2 pandemic. The possibility of coping with adolescents' risky behaviors was studied by Natali et al. [1]. They provided evidence that a single workshop combining psychoeducation and skills training, which lasted for only 3 h, could improve teachers' confidence and feelings of potential self-efficacy to respond to these behaviors. The authors evidenced the potential of online sessions to increase the efficacy of this intervention and to offer it to a larger number of teachers.

Cossu et al. [2] studied 55 adolescents presenting with suicidal ideation and/or suicide attempt. They used the Minnesota Multiphasic Personality Inventory—Adolescents, evidencing statistically significant differences between the two groups. Adolescents who attempted suicide had higher levels of relational difficulties, more risky behaviors, reduced aspirations for their future, and also a higher tendency to lie, as well as an increased use of repression. These data, although needing to be confirmed in larger prospective studies, offered a possible way to support strategies to stratify suicide attempt risk and, therefore, interventions.

This leads to the second group of papers, i.e., those studying what happened to children and adolescents during the SARS-CoV-2 pandemic. Pontillo et al. [3] compared the

Received: 26 January 2025
Accepted: 10 February 2025
Published: 12 February 2025

Citation: Chiappedi, M. Child and Adolescent Psychiatry: Unmet Needs Before and After the COVID-19 Pandemic. *Children* **2025**, *12*, 218. https://doi.org/10.3390/children12020218

number of inpatients followed in their Child and Adolescent Neuropsychiatry Unit before and during the pandemic in Italy. They showed a significant reduction in the number of inpatients during the peak of the pandemic and an increase in the following months, supporting the idea that "stay at home" orders had also reduced access to treatments. This is even more relevant because the authors also showed that mood disorders, non-suicidal self-injurious behaviors and suicidal ideation increased significantly during the "stay at home" period (also named "lockdown").

This increase in mood symptoms was confirmed by Shakarun et al. [4] in their study conducted in Saskatchewan, Canada. Their study involved 563 child–parent dyads and showed that hybrid learning, disrupted activities and increased screen time worsened mood alterations, as well as being part of an ethnic minority and living in smaller cities.

Pedrini et al. [5] studied in detail how changes in lifestyle habits due to the SARS-CoV-2 pandemic acted on anxiety in a group of adolescents. According to their regression analyses, increases in anxiety were associated with sleep problems, difficulties in reducing screen time and loneliness. Interestingly, their data seem to suggest a possible long-term effect of the pandemic and of the public health measures (including "stay at home" orders), both in changing lifestyle habits and their purpose and impact on one's mental health.

Akin and Sarrar [6] conducted a cross-sectional case–control study in Germany. Their data confirmed the presence of more atypical aspects of personality development, coupled with a less mature defense style and higher levels of psychodynamic conflicts in adolescents with mental health issues, leading to a number of disorders involving the body (somatoform symptoms, eating disorders, and alcohol use disorders) and/or mood (depressive disorders) and/or anxiety. During the pandemic period, these adolescents showed lower levels of conflict on the topic of autonomy (i.e., taking care of oneself versus being cared for by parents in a more or less passive way).

The role of parenting was studied by Facci et al. [7] in Italy during the pandemic period. They enrolled 136 mothers of preschool children for a survey and used multiple regression analyses to show an association between warmth and negative feelings on the one hand and positive parenting on the other; this was moderated by dismissing parental style toward children's emotions. These data support the importance of sharing emotions and feelings, although negative and difficult to think about, to improve childrens' development, even in times of crisis.

Working on even younger subjects, Richter et al. [8] studied infant regulatory problems, comparing their incidence in the pandemic versus the post-pandemic period in Germany. Crying/whining/sleeping problems and excessive crying had a higher prevalence in the post-pandemic period and were associated with less positive parenting behaviors in both mothers and fathers, this being partially mediated by parenting stress. This proved the importance of addressing infant mental health and to see parenting stress as a possible entry point for therapeutic interventions.

School closure in the context of "stay at home" orders was another disruption of children's and adolescents' "life routine". Berger et al. [9] studied the possibility for school mental health professionals in Australia to support and address the mental health needs of young people during the SARS-CoV-2 pandemic. These authors evidenced a number of relevant concerns regarding remote evaluations and counseling, both in terms of the reliability of assessment tools and of ethical issues.

These aspects are even more important in view of the findings of Lacombe et al. [10]. These authors interviewed 184 pre-adolescents using self-reported questionnaires. They found an increase in school burnout compared to what had been obtained in a similar survey in 2014. In this respect, the factors with the greatest impact were reduced confidence in the future, perceived stress, parental support and mathematics results; other significant factors were somatic symptoms and a negative and non-supportive classroom climate.

The Russo-Ukrainian war started in 2022 and provided an example of a non-health crisis, whose effects on adolescents were compared to those of the SARS-CoV-2 pandemic by Šalčiūnaitė-Nikonovė et al. [11]. These authors showed high levels of significant anxiety during both periods, although predictors were different (stress, problematic social media use and female gender for the pandemic period; stress, loneliness and lower self-efficacy for war period).

It is sad to have to face these kind of "natural experiments" [12]; as far as they are not avoidable by the best efforts of the scientific community, it is at least our duty to exploit them to improve our understanding and interventions.

Funding: This research received no external funding.

Conflicts of Interest: The author declares no conflicts of interest.

References

1. Natali, L.; Cardi, V.; Lunghi, M.; Ferrara, R.; Marconi, L.; Bottesi, G. The Acceptability of a Psychoeducation and Skill-Based Training for Carers and Teachers to Cope with Risky Behaviours in Adolescence. *Children* **2024**, *11*, 38. [CrossRef] [PubMed]
2. Cossu, G.; Vecchio, A.; Orlandi, M.; Casini, E.; Borgatti, R.; Mensi, M.; The Mondino Foundation Suicidality Research Group. Multiphasic Personality Assessment in a Case Series of Adolescent Patients with Suicidal Ideation and/or Attempts. *Children* **2023**, *10*, 1794. [CrossRef] [PubMed]
3. Pontillo, M.; Menghini, D.; Averna, R.; Labonia, M.; Lazzaro, G.; Tata, M.; Vicari, S. Psychiatric Hospitalization for Psychopathological Disorders and Self-Injurious Behaviors in Italian Children and Adolescents During COVID-19. *Children* **2023**, *10*, 1846. [CrossRef]
4. Shakurun, N.; Hinz, T.; Adeyinka, D.A.; Muhajarine, N. Unpacking the Mood States of Children and Youth in Saskatchewan, Canada, in the Context of the COVID-19 Pandemic: Insights from the "See Us, Hear Us 2.0" Study. *Children* **2025**, *12*, 79. [CrossRef] [PubMed]
5. Pedrini, L.; Meloni, S.; Dawson, J.; Geviti, A.; Bonfiglio, N.; Cattaneo, A.; Rossi, R. The Associations between Lifestyle Habits and Anxiety: A Prospective Study on Adolescents One Year after the Outbreak of the COVID-19 Pandemic. *Children* **2024**, *11*, 282. [CrossRef] [PubMed]
6. Akın, A.; Sarrar, L. Understanding Adolescent Mental Health in the COVID-19 Era: A Psychodynamic Approach. *Children* **2024**, *11*, 772. [CrossRef] [PubMed]
7. Facci, C.; Baroncelli, A.; Ciucci, E. The Role of Affects and Emotional Styles in the Relationship Between Parents and Preschool Children. *Children* **2024**, *11*, 1369. [CrossRef] [PubMed]
8. Richter, K.; Friedmann, A.; Mall, V.; Augustin, M. Infant Crying, Sleeping, and Feeding Problems in Times of Societal Crises: The Mediating Role of Parenting Stress on Parenting Behavior in Fathers and Mothers. *Children* **2024**, *11*, 1540. [CrossRef]
9. Berger, E.; Mackie, G.; Reupert, A.; Greenfeld, D.; Allen, K.; May, F.; Wurf, G.; Summers, D.; Morris, Z. The Experiences of Australian School Mental Health Professionals During COVID-19 Lockdowns. *Children* **2023**, *10*, 1157. [CrossRef] [PubMed]
10. Lacombe, N.; Hey, M.; Hofmann, V.; Pagnotta, C.; Squillaci, M. School Burnout after COVID-19, Prevalence and Role of Different Risk and Protective Factors in Preteen Students. *Children* **2023**, *10*, 823. [CrossRef]
11. Šalčiūnaitė-Nikonovė, L.; Žemaitaitytė, M.; Šmigelskas, K. Exploring the Prevalence and Predictors of Anxiety among Lithuanian Adolescents during Times of Crisis: A Cross-Sectional Study. *Children* **2024**, *11*, 32. [CrossRef] [PubMed]
12. Camporesi, A.; Vetrugno, L.; Buonsenso, D. COVID-19 and Children: Reflections after Three Years. *Children* **2024**, *11*, 10. [CrossRef] [PubMed]

children

MDPI

Article

The Acceptability of a Psychoeducation and Skill-Based Training for Carers and Teachers to Cope with Risky Behaviours in Adolescence

Ludovica Natali [1,2], Valentina Cardi [1,2,3], Marco Lunghi [1,4], Rosanna Ferrara [1], Linda Marconi [1] and Gioia Bottesi [1,2,*]

1 Centro di Ateneo Servizi Clinici Universitari Psicologici, University of Padova, 35131 Padova, Italy; ludovica.natali@phd.unipd.it (L.N.); valentina.cardi@unipd.it (V.C.); marco.lunghi@unipd.it (M.L.); rosanna.ferrara@unipd.it (R.F.); linda.marconi@unipd.it (L.M.)
2 Department of General Psychology, University of Padova, 35131 Padova, Italy
3 Department of Psychological Medicine, Institute of Psychiatry, Psychology and Neuroscience, King's College London, London SE5 8AF, UK
4 Department of Developmental Psychology and Socialisation, University of Padova, 35131 Padova, Italy
* Correspondence: gioia.bottesi@unipd.it

Abstract: Childhood and adolescence psychopathology is associated with an increased risk of psychological difficulties in adulthood. Early interventions for youth should provide carers and teachers with knowledge and skills to respond to adolescents' risky behaviours. This study evaluated the acceptability and effectiveness of a single 3-h workshop, combining psychoeducation and skills training to promote knowledge about, and confidence to address, adolescents' risky behaviours in carers and teachers of adolescents aged 10–14. Demographics and perceived self-efficacy in the parental or teaching role were collected at baseline using self-report questionnaires. Motivation and confidence to respond to adolescents' risky behaviours were measured before and after the workshop using motivational rulers. Participants provided written feedback about their experience about the workshop. Twenty-seven carers and 27 teachers attended the workshops. Teachers reported a significant increase in both importance ($p = 0.021$) and confidence ($p < 0.001$) to respond to risky behaviours following the workshop. This change was associated with baseline self-efficacy levels (importance: $p = 0.011$; confidence: $p = 0.002$). Carers also reported greater confidence to address risky behaviours following the workshop ($p = 0.002$). Participants found the contents and methods of the workshop highly acceptable. Online and multiple-session workshops might increase reach and effectiveness.

Keywords: psychoeducation; skills-training; adolescence; risky behaviours

Citation: Natali, L.; Cardi, V.; Lunghi, M.; Ferrara, R.; Marconi, L.; Bottesi, G. The Acceptability of a Psychoeducation and Skill-Based Training for Carers and Teachers to Cope with Risky Behaviours in Adolescence. *Children* **2024**, *11*, 38. https://doi.org/10.3390/children11010038

Academic Editor: Matteo Alessio Chiappedi

Received: 8 November 2023
Revised: 21 December 2023
Accepted: 26 December 2023
Published: 28 December 2023

1. Introduction

Adolescence is a phase characterized by heightened psychological vulnerability [1]. Approximately 8.8% of children and adolescents worldwide have been diagnosed with a mental disorder [2]. The increased vulnerability to mental health problems in adolescence is partly due to the numerous developmental challenges and tasks that young individuals face in this phase (e.g., search of their own identity, achieving autonomy) [3–5]. Furthermore, in recent years, worries and uncertainties about major events, such as the COVID-19 pandemic and climate change, have contributed to increased adolescent vulnerability to psychological distress [6–8]. Over time, if not addressed effectively, psychological difficulties early in life can lead to maladaptive outcomes in adulthood, including emotional and interpersonal difficulties, and poor mental health [9,10]. This posits the need for prevention programmes and early interventions to address youth psychological difficulties. A recently formulated roadmap to develop effective early interventions highlights the need for greater exchange of knowledge and skills between professionals and carers [11]. This is because significant others can influence the trajectory of mental health difficulties, either reinforcing these difficulties or providing valuable support to tackle them [12–14].

In recent years, an increasing number of studies have explored the effectiveness of single-session interventions (SSIs). SSIs can be defined as "specific, structured programs that intentionally involve just one visit or encounter with a clinic, provider, or program" [15]. SSIs include core components of comprehensive, evidence-based interventions and deliver them succinctly to enhance access and completion rates. As a result, appropriately targeted SSIs can offer a cost-effective supplement or alternative to standard care, particularly when considering that longer interventions do not necessarily equal more effective outcomes [16,17]. So far, single-session parent training has yielded positive outcomes, including improvements in parents' knowledge, well-being, and self-efficacy [18–20].

For this purpose, a single-session workshop for carers and teachers of adolescents aged 10–14 was developed, with the goal of providing them with knowledge and skills to identify and address at-risk behaviours. In particular, the workshop addressed topics such as developmental changes in adolescence and factors which might provide mental health risk or resilience. This study reports on carers' and teachers' outcomes of the workshops. The primary aim of the study was to assess the acceptability of the workshops (e.g., the degree of perceived involvement and satisfaction; strengths and weaknesses). Secondary aims were: (1) assessing the short-term impact of the workshops on participants' importance and confidence to identify and respond to adolescents' risky behaviours; (2) examining whether this change correlated with participants' self-efficacy at baseline. Studies on health behaviour indicate that motivation, as well as self-efficacy, mediate the impact of learning on behaviour change [21,22]. Accordingly, changes in motivation were considered as proximal indicators of behaviour change. The hypotheses were that the workshops would have led to greater importance and confidence to identify and respond to risky behaviours in adolescence compared to baseline. An additional goal was to examine the relationship between baseline levels of self-efficacy and changes in motivation.

2. Materials and Methods

2.1. Participants

The study targeted carers and teachers of adolescents aged 10–14 years. There were no exclusion criteria. Participants were recruited from estate (public) secondary schools in the northeast of Italy. All participants were initially informed about the workshop through informative posters and flyers made available in schools. Particular attention was given to the recruitment of families belonging to ethnic minorities. All carers who participated in the workshop were parents.

2.2. Intervention

The intervention consisted of a single 3-h workshop specifically designed for either teachers or carers (i.e., no joint teacher–parent workshops were organised). In total, four workshops were conducted (two for carers, two for teachers) from March 2023 until May 2023 (the project's duration was 17 months, starting on June 2022). Three expert psychologists (ML, LM, RF) specialized in clinical psychology, developmental psychology, and school psychology facilitated the workshops. All workshops were divided into two parts.

In the first part, participants were provided with psychoeducation about the changes and processes characterizing typical development during adolescence, with particular reference to neurobiological maturation, the development of cognitive and metacognitive abilities, the formation of one's identity, the salience of peers, and increased academic demands and self-regulation (e.g., [3–5,23]). Particular attention was given to how different factors, such as temperament, beliefs, parenting style, scholastic experiences, motivation, intelligence, and locus of control could shape the developmental trajectory (e.g., [24–26]).

In the second part of the workshop, participants were encouraged to learn by experience. They were asked to identify and respond to difficult at-risk behaviours by reading and answering questions about a series of illustrative vignettes.

2.3. Vignettes

Six illustrative vignettes were developed for the skills-training component of the workshop (three for carers and three for teachers). Each vignette described a scenario aimed at training participants in identifying and responding to risky behaviours in adolescents. For teachers, the vignettes covered the following themes: (1) learning difficulties; (2) ethnic minorities; (3) scholastic overachievement. For carers, the topics included: (1) autonomy and peer influence; (2) scholastic overachievement and risk of isolation; (3) scholastic demands. Examples of vignettes are shown in Table 1. After reading each vignette, participants were divided into groups and asked to answer the questions.

Table 1. Examples of vignettes and related questions.

Teachers: Vignette 3	
Luisa is 13 years old and she attends secondary school. She is a bright and high-achieving girl. She is well-integrated in her class and displays a positive attitude both towards her peers and teachers. Over the last few months, Luisa has avoided school at times when academic tests were due. Her goal is to succeed at school with very high marks. In order to achieve this goal, she believes it would be necessary to present herself well in high school. Despite receiving positive feedback from her teachers, Luisa often feels dissatisfied in herself. She tends to focus on little mistakes and perceives positive criticism in a negative way. She used to practice gymnastics at a competitive level. However, this year, she decided to give up this sport to be able to focus more on her studies. During parent–teacher meetings, teachers praise Luisa's excellent abilities and her behaviour in class. Her parents believes that Luisa behaves very well at home, too. She spends most of her time at home studying, paying attention to even the smallest details until she achieves the top level. Luisa's parents have noticed that she is worried about her academic career and that she suffers from stomach aches and headaches when a test is due. On a few occasions, they have allowed her to remain home and miss school.	How might Luisa be feeling? What behaviours are exhibited by Luisa? What does Luisa think about the situation? What perception might she have of herself? What factors have contributed to the situation and Luisa's experiences? If no action took place, what could happen? On which factors could the teachers potentially intervene to help Luisa in this difficult situation?
Carers: Vignette 1	
Angela is a second-grade student in secondary school. Growing up, she has always been very extroverted and willing to play sports, causing no issues or concerns for her parents. At the parent–teacher meetings, the teachers report that Angela is hardworking and achieving good academic results. They also mention that, during recess, Angela tends to isolate from the rest of the class, often spending most of the time sitting at her desk, looking at her phone. Following the teachers' feedback, Angela's parents begin to observe her behaviour at home. They notice that she no longer goes out with her friends during the weekend and that she spends most of her time online on her phone. After a few weeks, during a family dinner, Angela starts sharing anecdotes about her friends. Her parents realize that they are unfamiliar with many of the names mentioned by her. They ask Angela about these friends, how she met them, and who they are. Angela says that they are friends she met through social media, with whom she shares common experiences and interests. Angela also says that she no longer wishes to participate in sports since she has never enjoyed doing so and that she has no interest in spending time outside.	What is the event to pay attention to? How might Angela be feeling? What behaviours are exhibited by Angela? What does Angela think about the situation? What perception might she have of herself? What factors have contributed to the situation and Angela's experiences? If no action took place, what could happen? On which factors could carers potentially intervene to help Angela in this situation?

2.4. Measures

Before the workshop, participants completed a demographic questionnaire including questions on age, gender, nationality, first language, level of education, marital status, number of children, financial income, psychological difficulties, and medical illnesses. All variables are displayed in Table 2. Participants also completed the following measures:

Table 2. Participants' socio-demographic characteristics expressed as means (standard deviations) or frequencies (%).

Variables	Teachers		Carers	
	N	*M (SD)* or Frequency (%)	*N*	*M (SD)* or Frequency (%)
Age	28	50.64 (7.35) [min:33; max:64]	32	48.49 (5.54) [min:41; max:66]
Gender (Female vs. Male)	29	28 (96.55%)	32	27 (84.38%)
Nationality (Italian vs. Other)	29		32	
Italian		29 (100%)		30 (93.75%)
European		0 (0.00%)		1 (3.13%)
Extra-European		0 (0.00%)		1 (3.13%)
First Language (Italian vs. Other)	28	28 (100%)	32	30 (93.75%)
Level of Education	28		32	
Inferior to Diploma		0 (0.00%)		2 (6.25%)
Diploma		0 (0.00%)		5 (15.63%)
Bachelor's Degree		1 (3.57%)		1 (3.13%)
Master's Degree		21 (75.00%)		21 (65.63%)
Other title (e.g., conservatory, academy of fine arts)		4 (14.29%)		1 (3.13%)
Marital Status	27		32	
Single		3 (11.11%)		3 (9.38%)
Cohabiting		3 (11.11%)		6 (18.75%)
Married		16 (59.26%)		21 (65.63%)
Separated		3 (11.11%)		1 (3.13%)
Divorced		2 (7.41%)		0 (0.00%)
Widowed		0 (0.00%)		1 (3.13%)
Number of Children	25	1.32 (1.07)	32	2.00 (0.62)
Income (Euros)	22		27	
<15,000		2 (9.09%)		0 (0.00%)
15,000–29,000		12 (54.55%)		9 (33.33%)
30,000–55,000		7 (31.82%)		14 (51.85%)
56,000–100,000		1 (4.55%)		4 (14.82%)
Accommodation	26		30	
Owned house		21 (80.77%)		22 (73.33%)
Rented house		4 (15.39%)		6 (20.00%)
Other		0 (0.00%)		2 (6.67%)
Employment [1]	-		32	
Full-time worker		-		14 (43.75%)
Part-time worker		-		10 (31.25%)
Self-employed		-		6 (18.75%)
Homemaker		-		1 (3.13%)
Unemployed (actively seeking employment)		-		1 (3.13%)
Years of teaching [2]	29	18.21 (9.16) [min:1.00–max:36.00]	-	-
Hours of teaching (per week) [2]	28	16.43 (5.53)	-	-
Number of classes [2]	28	4.75 (4.07)	-	-
Students with special needs (Yes vs. No) [2]	26	24 (92.31%)	-	-
Psychological Difficulties (Yes vs. No)	27	3 (11.11%)	32	5 (15.63%)
Medical Illnesses (Yes vs. No)	27	4 (14.81%)	32	3 (9.38%)
DASS-21 total score	29	10.31 (7.15)	31	11.58 (10.24)
TSES-SF [2] total score	29	6.60 (1.01)	-	-
PSOC [1] total score	-	-	30	64.07 (9.41)

Notes. DASS-21 = Depression, Anxiety and Stress Scale—21; TSES = Teachers' Sense of Efficacy Scale-Short Form; PSOC = Parenting Sense of Competence. [1] Variables collected only for carers. [2] Variables collected only for teachers.

The Depression, Anxiety, Stress Scale-21 (DASS-21) [27,28] to measure carers' and teachers' psychological distress. Items are scored on a 4-point Likert scale, where higher scores indicate greater severity of the symptoms in the last week. The total score demonstrated excellent internal consistency in this study (Cronbach's Alpha for teachers and carers = 0.9).

The Short Form of the Teachers' Sense of Efficacy Scale (TSES-SF) [29,30] to assess teachers' self-efficacy. The scale comprises three subscales (i.e., teacher self-efficacy in student engagement, teacher self-efficacy in instructional strategies, and teacher self-efficacy in classroom management) and a total score. Items are scored on a 9-point Likert scale, where higher scores indicate greater self-efficacy. The total score demonstrated excellent internal consistency in this study (Cronbach's Alpha = 0.9).

The Parenting Sense of Competence (PSOC) [31] to assess carers' efficacy in the parenting role. The questionnaire comprises two subscales (i.e., satisfaction and efficiency) and a total score. Items are scored on a 6-point Likert scale, and higher score indicate greater parenting sense of efficacy. There is an Italian version of the PSOC [32] and it has been used in other studies [33,34]. The total score of this instrument demonstrated good internal consistency in this study (Cronbach's Alpha = 0.8).

The motivational ruler to assess participants' motivation and confidence to respond to risky behaviours in relation to the situation described in the vignette. Two questions (one for motivation and one for confidence) were designed for each vignette and rated on a Visual Analogue Scale (VAS) from 0 ("Not at all") to 10 ("Extremely").

Qualitative feedback form. This evaluation form consisted of 7 items designed to collect participants' feedback on the workshop's content and delivery method. Four items were scored on a 4-point Likert scale (0 = "Not at all"-3 = "Very much"). The questions evaluated the clarity of the objectives (i.e., "How clear were the workshop objectives?"), the usefulness of the workshop (i.e., "How useful was the workshop to you?"), the adequacy of the delivery methods (i.e., "In your opinion, were the workshop delivery methods adequate?"), and the level of participation/involvement (i.e., "How engaged did you feel?"). Three questions were open-ended and asked information about the strengths (i.e., "In your opinion, what were the strengths of the workshop?") and weaknesses of the workshop (i.e., "In your opinion, what were the weaknesses of the workshop?"), as well as if there would have been other topics participants would have liked to discuss (i.e., "Were there any topics you would have liked to discuss during the workshop? If so, which ones?").

2.5. Procedure

Schools interested in the workshops were identified through the collaboration with local parents and teachers organizations (i.e., the Veneto Regional Coordination of School Council Presidents) and the municipal administration of Conegliano. Representatives of each school provided the complete list of carers and teachers interested in attending the workshop. The workshops took place in the school premises. One week prior to the workshop, participants completed an informed consent form and the baseline assessment through the online platform Qualtrics. On the day of the workshop, before and after the session, participants were given the vignettes and asked to complete the motivational ruler. At the end of the workshop, participants also completed the qualitative feedback form.

Ethical approval was obtained by the Ethics Committee for the Psychological Research of the University of Padova (reference number: 5195). The study was conducted in accordance with the Declaration of Helsinki. Written informed consent was provided by all participants.

2.6. Data Analysis

Descriptive analyses were used to describe the socio-demographic characteristics of the participants and to evaluate the acceptability of the workshops.

Paired-samples Wilcoxon tests were calculated on the motivational rulers to assess the impact of the workshops on motivation and confidence to address risky behaviours. Global scores for confidence and importance were obtained by averaging participants' answers to the vignettes. Rank biserial correlations were used to estimate effect sizes (ESs). The coefficient was described as tiny (<0.5), very small (≥0.05 and <0.1), small (≥0.1 and <0.2), medium (≥0.2 and <0.3), large (≥0.3 and <0.4), and very large (≥0.4) [35].

Spearman's correlations were performed to examine whether changes in motivation and confidence to tackle risky behaviours throughout the workshop would be related to teachers' and carers' self-efficacy at baseline. To do so, delta scores for motivation and confidence were computed, with positive scores indicating an improvement.

All the above-mentioned analyses were carried out on JASP [36], and statistical significance was set at $p < 0.05$.

3. Results

3.1. Socio-Demographic Characteristics

Thirty-four teachers and 37 carers expressed interest in participating the workshop. However, the final sample included 29 teachers and 32 carers, i.e., those who completed at least the baseline questionnaire. The flow of participation in the study is described in Figure 1. Participants' socio-demographic characteristics are presented in Table 2.

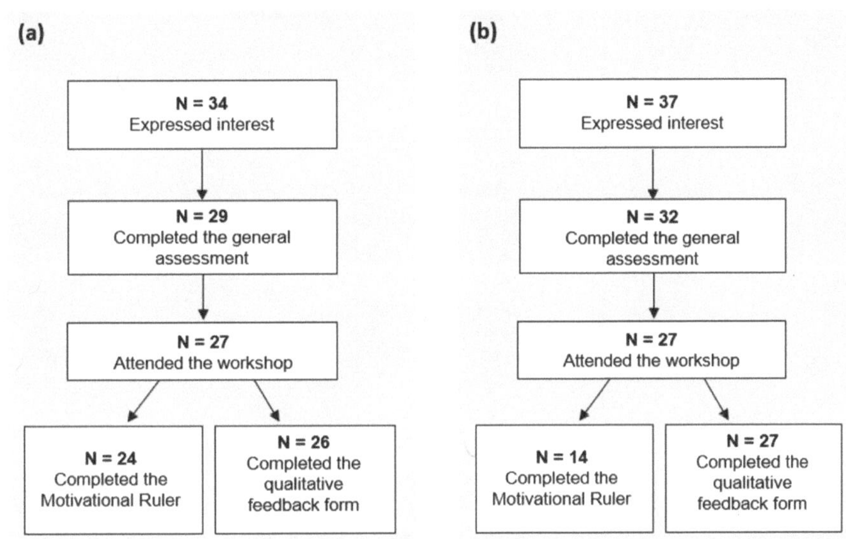

Figure 1. A CONSORT diagram describing the participation to the study of (**a**) teachers and (**b**) carers.

The mean age of teachers was 50.64 (7.35), while the mean age of carers (all parents) was 48.49 (5.54). Most of the sample comprised women (n = 55, 90.16%) and Italian nationals (n = 69, 96.72%). Most participants had a master's degree (n = 42, 70.00%), were married (n = 37, 62.71%), and were house owners (n = 43, 76.79%). Teachers reported an average number of children of 1.32 (1.07), while carers reported 2.00 (0.62). A slight difference was observed between teachers and carers with regards to income, with teachers reporting overall a smaller income. Only a small percentage of the sample reported experiencing psychological difficulties (n = 8, 13.56%) or medical conditions (n = 7, 11.86%).

3.2. Psychological Distress and Self-Efficacy

Most of the sample reported low levels of psychological distress (Teachers: n = 28, 96.55%; Carers: n = 29, 93.55%) with only a small minority reporting high levels (Teachers: n = 1, 3.45%; Carers: n = 2, 6.45%) [27].

Teachers reported lower self-efficacy compared to the normative Italian sample ([29]; N = 200; Mean = 7.02; SD = 1.45; t = −2.25, p = 0.03). Carers, from an observational perspective (since Italian normative data for comparison were unavailable), reported slightly higher self-efficacy compared to an Italian sample of mothers of children with typical development ([37]: $N = 240$; Mean = 60.47; SD = 10.85).

3.3. Workshop Acceptability

Overall, out of the 71 participants who initially expressed their interest in attending the workshop, 54 individuals (76.06%) attended the workshop, with 48 of them (67.61%) completing the motivational ruler and 53 (74.65%) completing the qualitative feedback form. Drop-outs were largely due to organizational difficulties (e.g., delayed start of the workshop, early closure of the school facilities). Descriptives on qualitative feedback are shown in Table 3.

Table 3. Qualitative feedback on the workshop.

Variables	Teachers		Carers	
	N	Frequency (%)	*N*	Frequency (%)
Clarity	26		27	
Not at all		0 (0.00%)		0 (0.00%)
Somewhat		2 (7.69%)		5 (18.52%)
Very		19 (73.08%)		14 (51.85%)
Extremely		5 (19.23%)		8 (29.63%)
Utility	26		27	
Not at all		0 (0.00%)		0 (0.00%)
Somewhat		6 (23.08%)		9 (33.33%)
Very		17 (65.39%)		14 (51.85%)
Extremely		3 (11.54%)		4 (14.82%)
Materials' suitability	26		26	
Not at all		0 (0.00%)		0 (0.00%)
Somewhat		4 (15.39%)		6 (23.06%)
Very		17 (65.39%)		13 (50.00%)
Extremely		5 (19.23%)		7 (26.93%)
Engagement	26		27	
Not at all		0 (0.00%)		0 (0.00%)
Somewhat		2 (7.69%)		2 (7.41%)
Very		21 (80.77%)		19 (70.37%)
Extremely		3 (11.54%)		6 (22.22%)
Strengths	26		23	
Skills-training/vignettes		20 (76.92%)		18 (78.26%)
Speakers' competence		9 (34.62%)		3 (13.04%)
Neurobiological insights		2 (7.69%)		0 (0.00%)
Weaknesses	26		23	
Lack of practical strategies		5 (19.23%)		1 (4.35%)
Short duration		7 (26.92%)		6 (26.09%)
Poor attendance		2 (7.69%)		6 (26.09%)
Suggested topics	26		23	
Specific disorders (e.g., eating disorders, learning disorders, disabilities)		2 (7.69%)		1 (4.35%)
Group interactions		8 (30.77%)		4 (17.39%)
New technologies and social media		2 (7.69%)		10 (43.48%)
School-family relationships		2 (7.69%)		3 (13.04%)

Overall, participants rated the workshop positively: the objectives were clear, the content was perceived as very helpful, they appreciated the delivery methods, and they felt very engaged. Participants felt that working in a group was one of the major strengths. They identified the one-off format of the workshop and the low attendance rate as weaknesses. They also expressed a desire for practical strategies to respond to problematic behaviours. Most teachers reported that they would be interested in learning more about group dynamics within the classroom and strategies for managing them. Carers were mostly interested in exploring the use of digital technologies and social media in adolescence.

3.4. Changes in Importance and Confidence to Address Risky Behaviours across the Workshop

Results of the Wilcoxon matched-pairs signed rank tests are reported in Table 4.

Table 4. Wilcoxon matched-pairs signed rank tests.

Teachers [1]					
Motivational ruler	Pre *M (SD)*	Post *M (SD)*	z	*p*-value	r_{rb}
Importance	7.81 (1.07)	8.15 (1.09)	2.053	0.021	0.524
Confidence	7.65 (0.89)	8.35 (1.02)	3.615	<0.001	0.900
Parents [2]					
Motivational ruler	Pre *M (SD)*	Post *M (SD)*	z	*p*-value	r_{rb}
Importance	9.10 (1.12)	9.29 (1.02)	0.700	0.241	0.306
Confidence	7.86 (1.22)	8.69 (1.00)	2.981	0.002	0.974

Notes. Importance and confidence to address adolescents' risky behaviours assessed using motivational rulers (Visual Analogue Scale from 0 = "Not at all" to 10 = "Extremely"). [1] N = 24. [2] N = 14.

A significant increase in teachers' importance ($p = 0.021$) and confidence ($p < 0.001$) to address risky behaviours was observed, with very large effect sizes. Likewise, carers reported a significant increase in their confidence to tackle risky behaviours after the workshop ($p = 0.002$), with a large effect size. No significant change in importance to respond to risky behaviours was observed in carers ($p = 0.241$).

3.5. Relationship between Changes in Motivation and Self-Efficacy

Teachers' baseline levels of self-efficacy were significantly associated with changes in teachers' importance ($r_s = -0.51$, $p = 0.011$, n = 24) and confidence ($r_s = -0.60$, $p = 0.002$, n = 24) to respond to risky behaviours. Specifically, greater improvements in importance and confidence to respond to risky behaviours were associated with lower teaching self-efficacy at baseline.

No significant correlations were found between carers' self-efficacy and their motivation ($r_s = -0.16$, $p = 0.626$, n = 12) and confidence ($r_s = -0.37$, $p = 0.234$, n = 12) to address risky behaviours.

4. Discussion

This article describes the preliminary findings of a workshop combining psychoeducation and skills training to promote knowledge about, and confidence to respond to, risky behaviours in adolescence. Findings indicated that the workshop was, overall, acceptable for participants, who reported higher levels of confidence to tackle risky behaviours in adolescence after attending the session. Teachers also reported greater levels of importance to tackle those issues following the workshop.

Overall, participants found the contents and methods of the workshops highly acceptable. They valued the skills-training component and the opportunity to engage in discussions with both peers and the group facilitators. Despite efforts to advertise the workshop widely to potential participants, only a limited number of individuals expressed interest in participating, and an even a smaller number completed the motivational ruler. Participants' feedback indicated that this was largely due to the timing of the workshop (late afternoon) and organizational issues related to the school facilities. This underscores the need for greater collaborative efforts to identify strategies to increase access. There is a significant demand for the development and implementation of mental health prevention, early detection, and support programs that target caregivers and teachers' needs. This is particularly important, as approximately 50% of all psychological disorders in adulthood begin by age 14 [38], and there has been a notable increase in prevalence rates, ranging from about 12% to more than 20%, of clinically severe anxiety and depression in youth cohorts following the COVID-19 pandemic [8]. Working together with carers and teachers to identify the best strategies to address their needs is a valuable opportunity to respond to this emergency.

The data on the short-term impact of the workshops on importance and confidence to respond to risky behaviours in adolescence were encouraging. Teachers reported a significant increase in both importance and confidence to respond to risky behaviours at the end of the workshop, and this change was associated with the baseline level of self-efficacy. It is possible that this type of intervention might be particularly beneficial for those with low levels of perceived self-efficacy. This is relevant when considering that teachers' self-efficacy is linked to students' outcomes (e.g., achievement and motivation) [39]. Carers reported a significant increase in their confidence to address risky behaviours, but not in importance. It is worth noting that carers, unlike teachers, expressed a high level of importance even before the workshop, as well as high self-efficacy, which could also explain why changes in importance and confidence to respond to risky behaviour were not associated with self-efficacy in this group. In this study, carers were all parents. Parents' attitudes towards their loved ones' mental health can vary depending on cultural, social, and individual factors. In many societies, there has been increasing awareness and recognition of the importance of mental health for adolescents, and parents are generally becoming more attuned to the mental health needs of their children, including teenagers [40]. The workshops in this study were attended on a volunteer basis and it is possible that a self-selected sample of parents participated. This adds to the poor generalizability of the findings, which is due to both the small sample size and its limited diversity (i.e., participants were predominantly Italian women, despite efforts to broaden recruitment to ethnic or other minorities; carers were all parents). Parent skills training might be perceived as "patronizing", and fear of stigmatization, shame and guilt might prevent them from taking part. The stigma associated with seeking mental health support and access services for children is heightened especially among parents from ethnic minorities and those with children having special needs [41–43]. One possible solution to enhance participation and inclusion might be transitioning to online workshops, which could overcome spatial constraints, provide greater flexibility, and allow greater privacy [44]. The online workshops could also be integrated into guided self-help interventions (GSH), to facilitate the use of self-help materials [45] that carers and teachers could use in their own time, beyond the time and space constraints of standard therapies. Within the study presented, short video clips were developed for those who could not attend the workshops.

Nonetheless, despite these limitations, the findings of this study add to the evidence that providing families with knowledge and skills to cope with their children's mental health difficulties is acceptable and associated with some benefit. For example, a recent systematic review of the literature on family support programs has demonstrated that these programs are effective in improving both caregivers' and children's mental health [46]. Similarly, the positive impact of the workshop found on teachers' confidence and importance to tackle difficult behaviours in adolescents corroborates previous findings on the efficacy of teachers' training to improve mental health literacy, reduce stigma, and increase confidence to offer help to students [47,48].

Future studies should include additional measures to assess a larger range of outcomes and identify groups who might especially benefit from interventions, such as the workshops tested in this study. These measures could, for example, assess teachers' and parents' stress (e.g., Teaching Stress Inventory [49], The Parental Stress Scale [50]), and adolescents' strengths and difficulties (e.g., Strengths and Difficulties Questionnaire [51]), and might be repeated over time, to follow up on changes in attitudes and behaviours. It also seems appropriate to evaluate social desirability and use semi-structured interviews to enable participants to provide more comprehensive feedback.

5. Conclusions

This study represents a first attempt to evaluate a novel SSI combining psychoeducation and skill training for carers and teachers, with the aim of enhancing their knowledge and confidence to address risky behaviours in adolescence. Preliminary findings are promising and suggest that the workshops are acceptable and helpful in increasing participants' confidence, particularly among those with lower self-efficacy.

The feasibility of (guided) self-help interventions for carers and teachers should be explored further, with a focus on increasing accessibility and sustaining benefits in the longer term. Working together with carers and teachers to identify the most effective intervention strategies is of paramount importance to meet these challenges.

Author Contributions: Conceptualization, V.C. and G.B.; Data curation, L.N.; Formal analysis, L.N.; Funding acquisition, V.C. and G.B.; Investigation, L.N., M.L., R.F. and L.M.; Methodology, L.N., M.L., R.F. and L.M.; Project administration, V.C. and G.B.; Supervision, V.C. and G.B.; Visualization, L.N.; Writing—original draft, L.N., V.C., M.L. and G.B.; Writing—review & editing, L.N., V.C., M.L., R.F., L.M. and G.B. All authors have read and agreed to the published version of the manuscript.

Funding: This research was supported by a grant from the University of Padova to the Centro di Ateneo Servizi Clinici Universitari Psicologici, within the scope of the "Bando per la presentazione di Progetti di Terza Missione—anno 2022".

Institutional Review Board Statement: The study was conducted in accordance with the Declaration of Helsinki and approved by the ethical committee of the University of Padova (reference number: 5195), the approval date was 12 March 2023.

Informed Consent Statement: Informed consent was obtained from all subjects involved in the study.

Data Availability Statement: The data presented in this study are available on request from the corresponding author. The data are not publicly available due to they report private information about participants.

Acknowledgments: The authors express their gratitude to Eva Bassi, Marco Ruffin, and Laura Tonon for their collaboration and assistance with the recruitment process. They also extend their thanks to Sabrina Cipolletta, Claudio Gentili, and Daniela Palomba for their support in identifying the project's objectives.

Conflicts of Interest: The authors declare no conflicts of interest.

References

1. McLaughlin, K.A.; King, K. Developmental Trajectories of Anxiety and Depression in Early Adolescence. *J. Abnorm. Child. Psychol.* **2015**, *43*, 311–323. [CrossRef] [PubMed]
2. Piao, J.; Huang, Y.; Han, C.; Li, Y.; Xu, Y.; Liu, Y.; He, X. Alarming Changes in the Global Burden of Mental Disorders in Children and Adolescents from 1990 to 2019: A Systematic Analysis for the Global Burden of Disease Study. *Eur. Child Adolesc. Psychiatry* **2022**, *31*, 1827–1845. [CrossRef] [PubMed]
3. Best, O.; Ban, S. Adolescence: Physical Changes and Neurological Development. *Br. J. Nurs.* **2021**, *30*, 272–275. [CrossRef] [PubMed]
4. Durkin, K. *Developmental Social Psychology: From Infancy to Old Age*; Blackwell Publishing: Oxford, UK, 1995.
5. Marcia, J.E. Identity in Adolescence. *Handb. Adolesc. Psychol.* **1980**, *9*, 159–187.
6. van Nieuwenhuizen, A.; Hudson, K.; Chen, X.; Hwong, A.R. The Effects of Climate Change on Child and Adolescent Mental Health: Clinical Considerations. *Curr. Psychiatry Rep.* **2021**, *23*, 88. [CrossRef] [PubMed]
7. Panchal, U.; Salazar de Pablo, G.; Franco, M.; Moreno, C.; Parellada, M.; Arango, C.; Fusar-Poli, P. The Impact of COVID-19 Lockdown on Child and Adolescent Mental Health: Systematic Review. *Eur. Child. Adolesc. Psychiatry* **2023**, *32*, 1151–1177. [CrossRef]
8. Racine, N.; McArthur, B.A.; Cooke, J.E.; Eirich, R.; Zhu, J.; Madigan, S. Global Prevalence of Depressive and Anxiety Symptoms in Children and Adolescents during COVID-19: A Meta-Analysis. *JAMA Pediatr.* **2021**, *175*, 1142–1150. [CrossRef]
9. Caspi, A.; Houts, R.M.; Ambler, A.; Danese, A.; Elliott, M.L.; Hariri, A.; Harrington, H.; Hogan, S.; Poulton, R.; Ramrakha, S. Longitudinal Assessment of Mental Health Disorders and Comorbidities across 4 Decades among Participants in the Dunedin Birth Cohort Study. *JAMA Netw. Open* **2020**, *3*, e203221. [CrossRef]
10. Laceulle, O.M.; Chung, J.M.; Vollebergh, W.A.M.; Ormel, J. The Wide-ranging Life Outcome Correlates of a General Psychopathology Factor in Adolescent Psychopathology. *Pers. Ment. Health* **2020**, *14*, 9–29. [CrossRef]

11. Uhlhaas, P.J.; Davey, C.G.; Mehta, U.M.; Shah, J.; Torous, J.; Allen, N.B.; Avenevoli, S.; Bella-Awusah, T.; Chanen, A.; Chen, E.Y.H. Towards a Youth Mental Health Paradigm: A Perspective and Roadmap. *Mol. Psychiatry* **2023**, *28*, 3171–3181. [CrossRef]
12. Griffith, J.M.; Crawford, C.M.; Oppenheimer, C.W.; Young, J.F.; Hankin, B.L. Parenting and Youth Onset of Depression across Three Years: Examining the Influence of Observed Parenting on Child and Adolescent Depressive Outcomes. *J. Abnorm. Child. Psychol.* **2019**, *47*, 1969–1980. [CrossRef] [PubMed]
13. Izon, E.; Berry, K.; Law, H.; French, P. Expressed Emotion (EE) in Families of Individuals at-Risk of Developing Psychosis: A Systematic Review. *Psychiatry Res.* **2018**, *270*, 661–672. [CrossRef] [PubMed]
14. Luebbe, A.M.; Bell, D.J. Positive and Negative Family Emotional Climate Differentially Predict Youth Anxiety and Depression via Distinct Affective Pathways. *J. Abnorm. Child. Psychol.* **2014**, *42*, 897–911. [CrossRef] [PubMed]
15. Schleider, J.L.; Beidas, R.S. Harnessing the Single-Session Intervention approach to promote scalable implementation of evidence-based practices in healthcare. *Front. Health Serv.* **2022**, *2*, 997406. [CrossRef] [PubMed]
16. Schleider, J.L.; Weisz, J.R. Little Treatments, Promising Effects? Meta-Analysis of Single-Session Interventions for Youth Psychiatric Problems. *J. Am. Acad. Child. Adolesc. Psychiatry* **2017**, *56*, 107–115. [CrossRef] [PubMed]
17. VanVoorhis, R.W.; Miller, K.L.; Miller, S.M. A Single-Session Intervention Designed to Promote Resilience for Parents of Children with Disabilities. *J. Child. Fam. Stud.* **2023**, *32*, 2406–2418. [CrossRef]
18. McConnell, M.; Closson, L.; Morse, B.; Wurster, H.; Flykt, M.; Sarche, M.; Biringen, Z. The "EA Brief": A Single Session of Parent Feedback and Coaching to Improve Emotional Attachment and Emotional Availability (EA). *Infant. Ment. Health J.* **2020**, *41*, 783–792. [CrossRef] [PubMed]
19. Lane, C.; Carson, V.; Morton, K.; Reno, K.; Wright, C.; Predy, M.; Naylor, P.-J. A Real-World Feasibility Study of the PLAYshop: A Brief Intervention to Facilitate Parent Engagement in Developing Their Child's Physical Literacy. *Pilot. Feasibility Stud.* **2021**, *7*, 113. [CrossRef]
20. Sommers-Flanagan, J. Single-Session Consultations for Parents: A Preliminary Investigation. *Fam. J.* **2007**, *15*, 24–29. [CrossRef]
21. Williams, B.W.; Kessler, H.A.; Williams, M. V Relationship among Knowledge Acquisition, Motivation to Change, and Self-efficacy in CME Participants. *J. Contin. Educ. Health Prof.* **2015**, *35*, S13–S21. [CrossRef]
22. Williams, B.W.; Kessler, H.A.; Williams, M. V Relationship among Practice Change, Motivation, and Self-efficacy. *J. Contin. Educ. Health Prof.* **2014**, *34*, S5–S10. [CrossRef]
23. Blasi, A.; Glodis, K. The Development of Identity. A Critical Analysis from the Perspective of the Self as Subject. *Dev. Rev.* **1995**, *15*, 404–433. [CrossRef]
24. Moe, A.; De Beni, R. Stile attributivo, motivazione ad apprendere ed atteggiamento strategico. Una rassegna. *Psicol. Clin. Dello Svilupp.* **2002**, *6*, 5–36.
25. Aunola, K.; Stattin, H.; Nurmi, J.-E. Parenting Styles and Adolescents' Achievement Strategies. *J. Adolesc.* **2000**, *23*, 205–222. [CrossRef]
26. Collins, W.A.; Maccoby, E.E.; Steinberg, L.; Hetherington, E.M.; Bornstein, M.H. Contemporary research on parenting: The case for nature and nurture. *Am. Psychol.* **2000**, *2*, 218–232. [CrossRef]
27. Bottesi, G.; Ghisi, M.; Altoè, G.; Conforti, E.; Melli, G.; Sica, C. The Italian Version of the Depression Anxiety Stress Scales-21: Factor Structure and Psychometric Properties on Community and Clinical Samples. *Compr. Psychiatry* **2015**, *60*, 170–181. [CrossRef]
28. Lovibond, P.F.; Lovibond, S.H. The Structure of Negative Emotional States: Comparison of the Depression Anxiety Stress Scales (DASS) with the Beck Depression and Anxiety Inventories. *Behav. Res. Ther.* **1995**, *33*, 335–343. [CrossRef]
29. Biasi, V.; Domenici, G.; Patrizi, N.; Capobianco, R. Teacher Self-Efficacy Scale (Scala Sull'auto-Efficacia Del Docente–SAED): Adattamento e Validazione in Italia. *J. Educ. Cult. Psychol. Stud.* **2014**, *10*, 485–509. [CrossRef]
30. Tschannen-Moran, M.; Hoy, A.W. Teacher Efficacy: Capturing an Elusive Construct. *Teach. Teach. Educ.* **2001**, *17*, 783–805. [CrossRef]
31. Johnston, C.; Mash, E.J. A Measure of Parenting Satisfaction and Efficacy. *J. Clin. Child. Psychol.* **1989**, *18*, 167–175. [CrossRef]
32. Vio, C.; Marzocchi, G.; Offredi, F. *Il Bambino Con Deficit Di Attenzione: Iperattività*; Diagnosi Psicologica e Formazione Dei Genitori; Edizioni Erickson: Trento, Italia, 1999.
33. Salomone, E.; Settanni, M.; McConachie, H.; Suma, K.; Ferrara, F.; Foletti, G.; Salandin, A.; WHO CST Team; Servili, C.; Adamson, L.B. Pilot Randomized Controlled Trial of the WHO Caregiver Skills Training in Public Health Services in Italy. *J. Autism Dev. Disord.* **2022**, *52*, 4286–4300. [CrossRef]
34. Uccelli, M.M.; Traversa, S.; Trojano, M.; Viterbo, R.G.; Ghezzi, A.; Signori, A. Lack of Information about Multiple Sclerosis in Children Can Impact Parents' Sense of Competency and Satisfaction within the Couple. *J. Neurol. Sci.* **2013**, *324*, 100–105. [CrossRef]
35. Funder, D.C.; Ozer, D.J. Evaluating Effect Size in Psychological Research: Sense and Nonsense. *Adv. Methods Pract. Psychol. Sci.* **2019**, *2*, 156–168. [CrossRef]
36. *JASP*, Version 0.16.3; Computer software. JASP Team: Amsterdam, The Netherlands, 2023.
37. Polizzi, C.; Giordano, G.; Burgio, S.; Lavanco, G.; Alesi, M. Maternal Competence, Maternal Burnout and Personality Traits in Italian Mothers after the First COVID-19 Lockdown. *Int. J. Environ. Res. Public. Health* **2022**, *19*, 9791. [CrossRef]
38. World Health Organization. Available online: https://www.who.int/news-room/fact-sheets/detail/adolescents-health-risks-and-solutions (accessed on 4 October 2023).
39. Hussain, M.S.; Khan, S.A.; Bidar, M.C. Self-Efficacy of Teachers: A Review of the Literature. *Jamshedpur Res. Rev.* **2022**, *1*, 110–116.

40. Kusaka, S.; Yamaguchi, S.; Foo, J.C.; Togo, F.; Sasaki, T. Mental Health Literacy Programs for Parents of Adolescents: A Systematic Review. *Front. Psychiatry* **2022**, *13*, 816508. [CrossRef]
41. Misra, S.; Jackson, V.W.; Chong, J.; Choe, K.; Tay, C.; Wong, J.; Yang, L.H. Systematic Review of Cultural Aspects of Stigma and Mental Illness among Racial and Ethnic Minority Groups in the United States: Implications for Interventions. *Am. J. Commun. Psychol.* **2021**, *68*, 486–512. [CrossRef]
42. Ali, A.; Hassiotis, A.; Strydom, A.; King, M. Self Stigma in People with Intellectual Disabilities and Courtesy Stigma in Family Carers: A Systematic Review. *Res. Dev. Disabil.* **2012**, *33*, 2122–2140. [CrossRef]
43. Turner, E.A.; Jensen-Doss, A.; Heffer, R.W. Ethnicity as a Moderator of How Parents' Attitudes and Perceived Stigma Influence Intentions to Seek Child Mental Health Services. *Cult. Divers. Ethn. Minor. Psychol.* **2015**, *21*, 613–618. [CrossRef]
44. Yap, M.B.H.; Mahtani, S.; Rapee, R.M.; Nicolas, C.; Lawrence, K.A.; Mackinnon, A.; Jorm, A.F. A Tailored Web-Based Intervention to Improve Parenting Risk and Protective Factors for Adolescent Depression and Anxiety Problems: Postintervention Findings from a Randomized Controlled Trial. *J. Med. Internet Res.* **2018**, *20*, e17. [CrossRef]
45. Cuijpers, P.; Schuurmans, J. Self-Help Interventions for Anxiety Disorders: An Overview. *Curr. Psychiatry Rep.* **2007**, *9*, 284–290. [CrossRef]
46. Kuhn, E.S.; Laird, R.D. Family support programs and adolescent mental health: Review of evidence. *Adolesc. Health Med. Ther.* **2014**, *5*, 127–142. [CrossRef]
47. Jorm, A.F.; Kitchener, B.A.; Sawyer, M.G.; Scales, H.; Cvetkovski, S. Mental health first aid training for high school teachers: A cluster randomized trial. *BMC Psychiatry* **2010**, *10*, 51. [CrossRef]
48. Santonja Ayuso, L.; Ruiz-Hontangas, A.; Cervantes, J.J.G.; Martínez, C.M.; Pons, E.G.; Pons, S.C.; Pejó, L.A.; Carmona-Simarro, J.V. The Promotion of Mental Health and Prevention of First-Episode Psychosis: A Pilot and Feasibility Non-Randomised Clinical Trial. *Int. J. Environ. Res. Public. Health* **2023**, *20*, 7087. [CrossRef]
49. Guidetti, G.; Viotti, S.; Badagliacca, R.; Converso, D. Looking for a Specific Measure for Assessing Sources of Stress among Teachers: A Proposal for an Italian Context. *Turk. Online J. Educ. Technol.* **2015**, *330*, 330–337.
50. Berry, J.O.; Jones, W.H. The Parental Stress Scale: Initial Psychometric Evidence. *J. Soc. Pers. Relat.* **1995**, *12*, 463–472. [CrossRef]
51. Tobia, V.; Marzocchi, G.M. The Strengths and Difficulties Questionnaire-Parents for Italian School-Aged Children: Psychometric Properties and Norms. *Child. Psychiatry Hum. Dev.* **2018**, *49*, 1–8. [CrossRef]

children

MDPI

Article

Multiphasic Personality Assessment in a Case Series of Adolescent Patients with Suicidal Ideation and/or Attempts

Giulia Cossu [1,†], Arianna Vecchio [2,†], Marika Orlandi [2], Erica Casini [2], Renato Borgatti [1,2,*], Martina Maria Mensi [2] and The Mondino Foundation Suicidality Research Group

1 Department of Brain and Behavioural Sciences, University of Pavia, 27100 Pavia, Italy
2 Child Neurology and Psychiatry Unit, IRCCS Mondino Foundation, 27100 Pavia, Italy
* Correspondence: renato.borgatti@mondino.it
† These authors contributed equally to this work.

Abstract: Suicide is an important public health issue. To examine the differences in personality characteristics between a group of adolescents with suicidal ideation (SI) and a group with a history of suicidal attempts (SA), we conducted a cross-sectional study. We enrolled 55 adolescents (51 females; 12–18 y.o.) who presented SI and/or SA. Using the Columbia Suicide Severity Rating Scale, we divided the sample into two groups: adolescents with SI and adolescents with SA. All participants filled in the Minnesota Multiphasic Personality Inventory—Adolescent (MMPI-A). Adolescents in the SA group had greater difficulties in social relations, risky behaviors, and more intense suicidal ideation compared to those in the SI group. Adolescents in the SA group scored higher in Omission, in the Lie Scale, the Conduct Problem Scale, the Less Aspirations Scale, the Repression Scale in the MMPI-A, and item 283 of the MAST compared to the other group. The results suggest that using the MMPI-A to assess certain features (e.g., tendency to lie, repression) may be helpful in identifying young people who are at high risk of suicide. However, further research is required to determine the effectiveness of using this instrument.

Keywords: adolescence; suicide attempt; suicide ideation; assessment; personality

Citation: Cossu, G.; Vecchio, A.; Orlandi, M.; Casini, E.; Borgatti, R.; Mensi, M.M.; The Mondino Foundation Suicidality Research Group. Multiphasic Personality Assessment in a Case Series of Adolescent Patients with Suicidal Ideation and/or Attempts. *Children* **2023**, *10*, 1794. https://doi.org/10.3390/children10111794

Academic Editors: Annio Posar and Brian Littlechild

Received: 8 October 2023
Revised: 3 November 2023
Accepted: 6 November 2023
Published: 8 November 2023

1. Introduction

Suicide is a rising phenomenon that affects individuals all around the world and is particularly prevalent in youth. According to research in 2020, suicide ranks as the second most common cause of death for those between the ages of 10 and 19 [1]. According to recent findings, social isolation brought on by the lockdown situation is one of the theories explaining the increased likelihood of suicidal thoughts during the COVID-19 epidemic [2–4]. In addition, suicide risk is correlated with several factors, including social isolation, as well as psychiatric, genetic, environmental, temperamental, and sociocultural aspects [5–7]. Several studies have documented an increased prevalence of suicidal attempts (SA) in relation to completed suicide among younger people compared to adults (ratio SA:S = 200:1), as well as in the female population compared to male peers of the same age [8]. These data derive from the hypothesis that women are more likely to be introspective and verbally disclose when they are uncomfortable. As a result, in these cases, the self-aggressive act would seem to have a communicative purpose, which is further supported by the fact that females are more likely to use non-violent methods. In contrast to women who mostly employ poisoning, especially from narcotics, men use blunt or gunshot wounds, hanging, and precipitation more commonly, which are more violent methods [9,10].

Major affective disorders are the psychiatric conditions most associated with suicide [10], particularly major depressive disorder [11,12]. An increased risk of suicide is also linked to psychotic symptoms [13,14], borderline personality disorder [15], and alcohol and substance abuse [16,17]. The assessment of suicide risk necessitates the consideration of previous suicide attempts [18], the existence of disadvantageous environmental situations [19], including the premature death of a parent [20], the suicide of a parent [21], and the presence of scholastic stressors [22], as these aspects constitute risk factors for a potential suicide attempt [23].

According to previous studies, suicidality and perfectionism seem to be related [24]. However, other stressful life events, such as being bullied or the existence of other personality characteristics, including neuroticism, anxiety, extraversion, and depression, appear to mitigate this factor [25]. As already established, there are several risk factors for suicide. Some of these have been the subject of more extensive research than others. These include personality traits, although prior studies advise assessing them in order to help physicians customize the most effective early intervention, especially for children and adolescents [26]. Several researchers found that certain personality traits and characteristics, such as borderline personality traits, seem to predispose people to suicide, neuroticism, psychoticism, and extraversion are other significant traits that are more closely linked to suicidality. Moreover, the literature highlighted that individuals who exhibit perfectionism, interpersonal dependence, novelty seeking, impulsivity, pessimism, nonconformity, low self-esteem, feelings of inferiority, hopelessness, and self-criticism are more likely to have suicidal thoughts and commit suicide [23,27–33]. According to a study by Beautrais and colleagues, there is a link between suicidal behavior and personality characteristics such as neuroticism and introversion, also expressed through withdrawal and social isolation [34].

The present study aimed to examine which personality features are more associated with suicidal ideation and behavior in the adolescent population and the differences between these two groups. We would expect significant differences between the two groups in MMPI-A subscales, in the validity subscales and in specific items of the D Scale.

2. Materials and Methods

2.1. Participants

We recruited 55 adolescents aged 12–18 years (extremes included) from April 2020 to July 2022. All patients who had simply experienced suicidal ideation (SI) or who had a history of non-fatal suicide attempts or interrupted ones (SA) were admitted to the Child Neuropsychiatry unit as either inpatients or outpatients. Using the Columbia Suicide Severity Rating Scale (C-SSRS) [35], we divided the sample into two different groups: SI and SA. Patients with intellectual disabilities or with an insufficient level of comprehension of the Italian language were deemed not eligible for the study. We used the Wechsler Intelligence Scale for Children-Fourth Edition (WISC-IV) [36] or the Wechsler Adult Intelligence Scale-Fourth Edition (WAIS-IV) [37], depending on age, to assess the presence of an intellectual disability. Figure 1 shows the study sample flowchart.

2.2. Ethics

This clinical cross-sectional study was realized according to the REporting of studies Conducted using Observational Routinely collected health Data (RECORD) (see Supplementary Material). The Ethics Committee of Policlinico San Matteo in Pavia, Italy, approved this study, which was conducted following the declaration of Helsinki (1964) and its later amendment. Patients gave their consent to participate in the study, and their parents signed the written informed consent. They had the possibility to withdraw from the study at any time without explaining. We anonymized data. The dataset is available in the Zenodo repository [38].

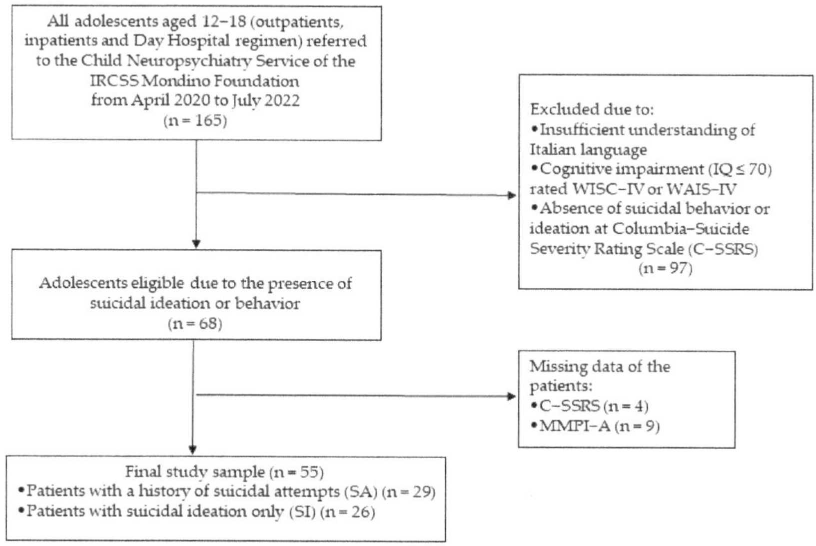

Figure 1. Study population flowchart [author's own processing].

2.3. Procedures and Measures

A child neuropsychiatrist collected sociodemographic and anamnestic data of the patients and verified that the patients had no intellectual disability and sufficient comprehension of the Italian language. We administered the Columbia Suicide Severity Rating Scale (C-SSRS) to assess the presence and the severity of SI and/or SA in the last 6 months. The C-SSRS is a semi-structured interview to evaluate the patient's severity of suicidal ideation. The instrument is very sensitive and permits the detection of the presence of suicidal ideation and suicidal behavior, as well as the frequency and gravity thereof. The suicidal ideation spectrum is taken into account by the scale, which goes from "wish to be dead 'to' active suicidal ideation with specific plans, intents, and behaviors".

Thereafter, a neuropsychiatrist or psychologist compiled the Children Global Assessment Scale (CGAS) [39] to assess the subject's psychosocial and work-related functioning. The CGAS is a 100-point rating scale composed of a continuum that ranges from mental health (91–100 = Superior functioning) to a serious mental disorder with a high risk of death (1–10 = need constant supervision). In addition, the participants compiled the Minnesota Multiphasic Personality Inventory—Adolescent (MMPI-A) [40], a questionnaire used to assess personality in adolescence. The MMPI-A consists of 478 true–false items and presents different scales:

- six validity scales: Lie (L), Variable Response Inconsistency (VRIN), True Response Inconsistency (TRIN), Infrequency 1(F1), Infrequency 2 (F2), Infrequency (F), Correction (K), Cannot Say (?);
- ten clinical scales: Hypochondria (Hs), Depression (D), Hysteria (Hy), Psychopathic deviation (Pd), Masculinity/Femininity (Mf), Paranoia (Pa), Psychasthenia (Pt), Schizophrenia (Sc), Hypomania (Ma), Social Introversion (Si);
- 15 content scales: Anxiety (A-anx), Obsessiveness (A-obs), Depression (A-dep), Health Concerns (A-hea), Alienation (A-aln), Bizarre Mentation (A-biz), Anger (A-ang), Cynicism (A-cyn), Conduct Problems(A-con), Low Self-Esteem (A-lse), Low Aspirations (A-las), Social Discomfort (A-sod), Family Problems (A-fam), School Problems(A-sch), Negative Treatment Indicators (A-trt);
- six supplementary scales: MacAndrew-Revised (MAC-R), Alcohol/Drug Problem Acknowledgement (ACK), Alcohol/Drug Problem Proneness (PRO), Immaturity (IMM), Anxiety(A), Repression(R).

Filling in the MMPI-A requires a lot of concentration, motivation, and good comprehension skills. Many adolescents compiled the questionnaire with pleasure because they can unveil personal characteristics and at least talk about themselves and their difficulties with clinicians; for many others, the test could be too demanding. To avoid invalid protocols or a lack of motivation, clinicians and psychologists were available for adolescents to help them read the sentences and better explain their meaning (while still adhering to the rules of administration). Moreover, after the assessment, the referring clinician took time to discuss the results with every patient and their expectations for the following treatment.

2.4. Statistical Analyses

We determined the sample size with G*Power [41,42] with an effect size of 0.7 and a power of 0.8. Descriptive analyses of the sample, which included demographic and clinical characteristics, were initially performed on the total sample, and subsequently, on the SA and SI groups separately. The analyses included means and standard deviations (SD) for continuous variables and absolute and relative frequencies for categorical variables. A comparison between the two groups was performed with an independent t-test for the numerical variables and with the Chi-square test for the categorical variables. We applied Fisher's correction as the number of cases was small. Statistical significance was expressed as $p < 0.05$. All statistical analyses were performed with IBM SPSS version 27.0 [43].

3. Results

The sample was formed of 55 participants with SI and/or SA. Table 1 shows the sociodemographic and anamnestic data for the total sample and for the two groups. With the exception of risky behaviors ($p = 0.027$) and social interactions ($p = 0.045$), the two groups were homogeneous. A comparison was made between the two groups concerning psychosocial functioning through the CGAS, from which no significant differences emerged ($M_{SA} = 48.81$, $SD_{SA} = 10.77$; $M_{SI} = 51$, $SD_{SI} = 10.5$). Regarding suicidal ideation, there were significant differences between the two groups ($p = 0.004$) in intensity but not in frequency. Suicidal ideation was more severe in the SA group than in the SI group. Table 2 shows the differences between the SI and SA groups.

Table 1. Sociodemographic and anamnestic data in the total sample and the two subgroups (author's own processing).

Variable: Number of Subjects and Percentage Indicated		Total (n = 55)	SI (n = 26)	SA (n = 29)
Gender	Male	4 (7.3)	3 (11.5)	1 (3.5)
	Female	51 (92.7)	23 (88.5)	28 (96.5)
Ethnicity	Caucasian	47 (85.5)	20 (76.9)	27 (93.1)
	African	3 (5.5)	2 (7.7)	1 (3.6)
	Latin (South America)	2 (3.6)	1 (3.8)	1 (3.6)
	Mixed	2 (3.6)	2 (7.7)	0
	Asian	1 (1.8)	1 (3.8)	0
Siblings	Only child	9 (16.4)	4 (15.4)	5 (17.2)
	Major	15 (27.2)	7 (26.9)	9 (31.1)
	Minor	25 (45.5)	11 (42.3)	13 (44.8)
	Major and minor	6 (10.9)	4 (15.4)	2 (6.9)
Parents	Married	31 (58.1)	15 (57.7)	16 (57.1)
	Divorced	23 (41.9)	11 (42.3)	12 (42.9)
SES	Low	5 (11.9)	4 (23.5)	1 (4)
	Medium-low	3 (7.1)	0	3 (12)
	Medium	13 (30.9)	5 (29.4)	8 (32)
	Medium-high	13 (30.9)	5 (29.4)	8 (32)
	High	8 (19.2)	3 (17.7)	5 (20)

Table 1. *Cont.*

Variable: Number of Subjects and Percentage Indicated		Total (n = 55)	SI (n = 26)	SA (n = 29)
Previous consultations	Absent	12 (22.2)	5 (19.3)	7 (25)
	Present	42 (77.8)	21 (80.7)	21 (75)
Social relationships	Social withdrawal	6 (10.9)	0	6 (20.7)
	Poor relations	30 (54.5)	16 (61.5)	14 (48.3)
	Adequate relations	19 (34.6)	10 (38.5)	9 (31.0)
Familiarity	Depression (sub-threshold)	19 (34.5)	9 (34.6)	10 (34.5)
	Depression (supra-threshold)	2 (3.6)	0	2 (6.9)
	Bipolar disorder (sub-threshold)	1 (1.8)	1 (3.8)	0
	Anxiety (sub-threshold)	9 (16.4)	4 (15.4)	5 (17.2)
	Substance abuse	2 (3.6)	2 (7.7)	0
	Eating behavior disorder	4 (7.3)	1 (3.8)	3 (10.3)
	Opp. Defiant behavior disorder	1 (1.8)	1 (3.8)	0
	Personality disorder (sub-threshold)	1 (1.8)	1 (3.8)	0
	Other	23 (41.8)	12 (46.2)	11 (37.9)
Risky behaviors	Substance use, self-harm, SA	40 (72.7)	15 (57.7)	25 (86.2)
Academic performance	Poor	9 (16.3)	3 (11.5)	6 (20.7)
	Sufficient	14 (25.5)	5 (19.3)	9 (31.1)
	Good	15 (27.7)	8 (30.8)	7 (24.1)
	Excellent	12 (21.8)	7 (26.9)	5 (17.2)
	School withdrawal	5 (9.1)	3 (11.5)	2 (6.9)
Access modes	Outpatient clinic	13 (23.6)	5 (19.3)	8 (27.6)
	Day hospital	1 (1.8)	1 (3.8)	0
	Hospitalization	41 (74.5)	20 (76.9)	21 (72.4)
Diagnosis	Neurodevelopmental disorder	1 (1.8)	0	1 (3.5)
	Schizophrenia/psychotic disorders	13 (23.6)	4 (15.4)	9 (31.0)
	Bipolar and related	2 (3.6)	2 (7.7)	0
	Depression	39 (70.9)	18 (69.2)	21 (72.4)
	Anxiety disorders	18 (32.7)	10 (38.5)	8 (27.6)
	DOC and related	3 (5.5)	1 (3.8)	1 (3.5)
	Eating disorders	15 (27.3)	9 (34.6)	6 (20.7)
	Personality disorders	34 (61.8)	7 (26.9)	14 (48.2)
	Substance use disorders	1 (1.8)	1 (3.8)	1 (3.5)
	Other	12 (21.8)	5 (19.2)	7 (24.1)
Previous psychotherapy	Absent	31 (56.3)	14 (53.8)	17 (58.6)
	Present	24 (43.7)	12 (46.2)	12 (41.4)
Pharmacotherapy prescribed	Absent	10 (20.0)	6 (23.1)	4 (14.8)
	Present	43 (80.0)	20 (76.9)	23 (85.2)
Type of medication	Antipsychotic	26 (47.2)	9 (32.1)	17 (60.7)
	Antidepressant	20 (36.4)	9 (32.1)	11 (39.3)
	Anxiolytic	30 (54.5)	12 (48)	18 (64.3)
	Mood stabilizer	4 (7.3)	1 (4)	3 (10.7)
CGAS	-	49.9	51	48.81
IQ	-	105.01	104. 52	105.48

3.1. Differences between the SI and SA Groups in the Validity Scales of the MMPI-A

We compared SI and SA groups with the number of MMPI-A protocols with 30 or more omissions, thus considered invalid. The analyses showed that all the SI group questionnaires were valid (n = 26), while in the SA group, 4 questionnaires were invalid. However, by performing the Chi-squared test with Fisher's correction, emerged no significant differences between the two groups ($p = 0.113$). When compared to the SA group, the SI group often responded to fewer questions (M_{SI} = 5.9; M_{SA} = 2.2; $p < 0.001$), and the difference between the groups was statistically significant; results showed a tendency in the SA group to have higher scores on the L (Lie) Scale ($p = 0.04$) compared to the SI group.

Table 2. Differences between the SI and SA groups in MMPI-A (author's own processing).

	SI (n = 26)		SA (n = 25)			
	M	SD	M	SD	t	*p*
Omissions	5.9	5.22	2.2	4.06	2.8	<0.001 **
VRIN	54.2	11.05	55.64	7.55	−0.53	0.299
TRIN-V	58.9	6.11	58.64	8.06	0.103	0.459
F1	72.7	15.68	73.84	15.25	−0.256	0.40
F2	70.5	16.91	75	14.88	−0.998	0.162
F	74.7	16.80	78.12	15.74	0.743	0.231
L	46.3	8.27	50.36	8.50	−1.743	0.044 *
K	45.62	7.45	47.08	6.70	−0.737	0.232
F-K	8.58	10.32	9.04	8.10	−0.178	0.43
Hs	69.92	11.23	68	12.65	0.568	0.286
D	79.88	9.77	79.8	11.49	0.027	0.489
Hy	68.88	11.98	68.88	12.01	0	0.5
Pd	66.60	10.81	68	11.27	−0.448	0.328
Mf	50.52	10.60	44.44	8.46	2.242	0.015 *
Pa	71.72	11.64	72.2	11.31	−0.148	0.442
Pt	71.20	9.67	70.4	10.01	0.287	0.388
Sc	75.72	13.47	73.12	13.43	0.683	0.249
Ma	52.08	8.60	51.68	10.87	0.144	0.443
Si	68.2	8.62	68.68	8.83	−0.195	0.423
ANX	70.4	11.655	72.6	8.836	−0.752	0.228
OBS	61	14.609	60.8	8.912	0.058	0.477
DEP	71.52	10.042	76.52	9.592	−1.8	0.39
HEA	69	11.836	66.8	13.793	0.61	0.274
ALN	73.56	15.314	70.56	12.484	0.759	0.226
BIZ	63.68	12.841	63.64	15.256	0.01	0.496
ANG	55.32	10.152	51.4	8.529	1.478	0.073
CYN	51	11.8	50.56	11.637	0.133	0.447
CON	57.76	15.621	50.52	8.945	2.011	0.026 *
LSE	70.44	14.897	69.88	10.365	0.154	0.439
LAS	66.12	12.862	59.2	12.203	1.951	0.028 *
SOD	75.84	10.703	76.76	13.767	−0.264	0.397
FAM	62.32	13.009	60.6	14.6	0.44	0.331
SCH	64.72	12.023	62.64	11.485	0.625	0.267
TRT	72.68	21.075	70.48	16.83	0.408	0.343
MAC/R	51.08	9.648	51.4	11.188	−0.108	0.457

Table 2. *Cont.*

	SI (n = 26)		SA (n = 25)			
	M	SD	M	SD	t	*p*
ACK	65.8	15.538	61.08	16.036	1.057	0.148
PRO	54.68	12.779	53.36	11.608	0.382	0.352
IMM	68.24	13.116	64.8	10.583	1.021	0.156
A	65.08	8.441	66.04	5.827	−0.468	0.321
R	57.64	10.169	62.6	8.602	−1.862	0.034 *
Item 68	0.77	0.43	0.92	0.282	−1.42	0.078
Item 71	0.5	0.51	0.36	0.49	0.999	0.161
Item 124	0.8	0.408	0.92	0.282	−1.159	0.125
Item 177	0.85	0.368	0.84	0.374	0.059	0.477
Item 283	0.54	0.508	0.88	0.332	−2.829	0.003 *
Item 347	0.84	0.374	0.87	0.338	−0.343	0.366
Item 371	0.81	0.402	0.8	0.408	0.068	0.473
Item 372	0.81	0.402	0.76	0.436	0.406	0.343
Item 399	0.77	0.43	0.84	0.374	−0.626	0.267

Significance: * $p < 0.05$, ** $p < 0.001$.

3.2. Differences between the SI and SA Groups in the Clinical Scales of the MMPI-A

Regarding the clinical scales, we found that the two groups presented statistically significant differences concerning the Masculinity-femininity scale (Mf) ($p = 0.015$). The SI group presented higher scores than the SA group.

3.3. Differences between the SI and SA Groups in the Content Scales of the MMPI-A

The results of the analyses revealed a statistically significant difference in two of the content scales of the MMPI-A questionnaire. We found an elevation in the Conduct Problems (CON) Scale ($p = 0.026$) and in the Low Aspirations (LAS) Scale ($p = 0.028$) in subjects belonging to the SA group.

3.4. Differences between the SI and SA Groups in the Supplementary Scales of the MMPI-A

The Repression (R) Scale, which has higher scores in the SA group than the SI group ($p = 0.034$), was the only additional scale where a statistically significant difference between the two groups was found.

3.5. Difference between the SI and SA Groups about Specific Items of the MMPI-A Causing an Elevation of the D Scale (Depression)

The analyses performed showed that the SI group and the SA group differed significantly in their responses to item 283, "Almost always I wish I was dead" ($p = 0.003$). It emerged that the subjects belonging to the SA group tended to give more affirmative answers to this item, with a mean score of 0.88, in contrast to the SI group, which had a mean score of 0.54.

4. Discussion

The present study aimed to investigate whether there were any differences in the validity, clinical, content, and additional scales of the MMPI-A between subjects who had experienced suicidal ideation only (SI) and those who had a history of non-fatal suicidal behaviors (SA). To the best of our knowledge, there are no further studies in the literature that investigate this issue in a sample of adolescents. Qualitative analyses in our study indicated no significant differences between the SI and SA groups in terms of familiarity with psychiatric disorders, diagnoses, and academic achievements. As the two groups presented homogenous features, they were considered comparable. There seems to be an association between suicidal behavior and self-directed life-threatening behaviors, such as self-injury and alcohol/substance abuse. According to earlier research, patients who attempt suicide and exhibit a preponderance of internalizing symptoms may turn to drugs as a coping mechanism for depressive symptoms [16,17,44]. In addition, in our study, we also found scores that may indicate the possibility of a correlation between suicidal behavior and social withdrawal. The analyses of our sample indicated that the SA group had more relational problems and social isolation. Our results confirm the findings of a recent review, which indicated that social isolation may be a predictor of SA in all ages and genders [45]. Adolescents who report a lack of social support and feelings of isolation may self-harm and attempt suicide more frequently. Therefore, we can state that they show a greater tendency to social withdrawal and violent behavior, which is more introjected and directed towards themselves rather than others. Piotti and colleagues have explained this hypothesis, supporting that the tendency of adolescents who experience suicidal behaviors to dissimulate their own suffering exists. Pietropolli Charmet's colleagues pointed out how our culture is characterized by the achievement of perfection and success, which "does not accept the experiences of failure, limitations, and dependence on others". In this context, people seem to wear a sort of "mask" that might fail during moments of crisis and expose their own vulnerabilities [46]. As the literature pointed out, in a society focused on achievement and success, the possibility of failure is minimized, and the fear of humiliation is a risk factor for SA [47,48].

These subjects do not manifest their suffering to others, which is instead directed towards themselves. Considering this aspect, Pietropolli Charmet's theory could explain the tendency of SA adolescents to self-procure physical pain through self-injury, hiding their suffering, and showing their competence to the outside world. Even the use of drugs can lead to long-term physical suffering. If initially the use of substances can be considered as a form of self-care, then it might become abuse, causing abstinence symptoms, leading in some cases to death.

According to the findings of our study on SA teenagers' propensity to conceal their pain, this group of patients may be more inclined to deny feelings of sadness and show a lack of sincerity towards others concerning their personal internal states. This is in line with the literature that shows low agreement between self-reports and clinician assessments regarding SI [49], with high rates of false positives registered in self-report measures compared to clinical assessment [50]. Moreover, previous studies reported different agreements between the family report and clinical assessment [51]. However, other studies report that participants feel more comfortable disclosing information on suicide-related topics through the relative anonymity of a self-rating scale [52,53].

A final difference emerged in our research between the two groups: the SA group reported lower scores on the masculinity–femininity (Mf) scale than the SI subjects. This difference shows that compared to persons exhibiting only SI, female subjects with SA may be less likely to exhibit stereotypically masculine interests and, thus, a lesser inclination to gender reversal. This result is in contrast to other research that suggested the prevalence of "psycho-sexual misfits" based on high Mf scale scores in the SA group; however, those studies focused on adult populations. [54]. We can also hypothesize that given that almost all the male participants in our study belonged to the SI group and had supra-threshold scores in the Mf scale, this difference could also be due to the unusual presence of stereotypically female interests in our sample males. It would be useful to investigate these aspects within the adolescent population, also considering how the concept of "correspondence" between one's biological gender and certain personological or behavioral characteristics has changed over time. Furthermore, the evolution of adolescents and their role within modern society should be taken into consideration [55,56].

Finally, concerning the intensity of suicidal ideation, in the present study, the differences in the scores on item 283—"Almost always I wish I was dead"—corroborate the results of several studies which showed that thoughts about death, wishing to be dead, thoughts of suicide, and suicide plans were significantly higher among adolescents with a history of SA [18]. The results from our study are consistent with the literature [18,57], which suggests that a substantial correlation exists between ideation and attempt, and that the existence of profound suicidal ideation and a prior suicidal attempt would be significant predictors of a subsequent attempt at suicide.

5. Conclusions

These findings may be useful in clinical practice to identify individuals who are more likely to attempt again; especially it seems important to pay attention to the subject's relational context. Social isolation may be a factor that increases the presence of suicidal behavior, as well as the presence of self-directed harming behavior. The presence of an externalizing condition is one feature that seems to be less associated with suicide; releasing anger outwardly may lessen the chance of engaging in a self-damaging act. In conclusion, despite the limits of this study due to the presence of a small sample of subjects, the lack of a control group, and the use of a self-administered and long questionnaire, the results of our study encourage the prosecution of research. In fact, at the moment, the small number of participants limits the generalizability of results, but we are enrolling more participants in order to do that. Furthermore, the MMPI-A could be useful in the assessment of adolescents at risk of suicidal behavior, analyzing personality aspects that are supposed to differentiate those who have already made an attempt from those who have not. The study we proposed is innovative, as there are no further studies in the literature that investigate differences in certain aspects of personality using the MMPI-A in a population of adolescents with SI and behavior. However, since this is a preliminary study, it also lacks longitudinal data. So, our research team is enrolling more participants to have a larger sample, continuing to monitor and collect data on the effects that the pandemic has had on this group of people [4,58], looking into potential risk factors for gender dysphoria [56], and including a control group. It may be possible to use the MMPI-A questionnaire as a preventive measure of the risk of suicide attempts in adolescents if it is administered in certain contexts (e.g., schools) together with additional questionnaires that can easily assess the presence of suicidal ideation, such as the Multi-Attitude Suicide Tendency Scale (MAST) [59]. We can hypothesize that by administering this questionnaire, it will be possible to determine which adolescents are most likely to have attempted suicide and to evaluate the presence of these characteristics in subjects who already present SI. This will allow us to consider the possibility of an intervention to prevent the act. We have found that certain personality traits, such as repression and a tendency to lie, are associated with the risk of suicide attempts in adolescents. Moreover, those results may represent the first step for future longitudinal studies investigating the interaction between clinical outcomes and the potentiality of MMPI-A as an efficient screening and preventive instrument.

Supplementary Materials: The following supporting information can be downloaded at: https://www.mdpi.com/article/10.3390/children10111794/s1. Ref. [60] is cited in the Supplementary Materials.

Author Contributions: Conceptualization, M.M.M.; data curation, G.C., A.V. and M.O.; formal analysis, E.C.; investigation, G.C., A.V. and M.O.; project administration, G.C., A.V. and M.O.; supervision, R.B. and M.M.M.; writing—original draft, G.C. and A.V.; writing—review and editing, G.C., A.V. and M.O. The members of The Mondino Foundation Suicidality Research Group contributed to data curation, supervision, and project administration. All authors have read and agreed to the published version of the manuscript.

Funding: The present study was supported by the Italian Ministry of Health (Ricerca Corrente 2022–2024).

Institutional Review Board Statement: The present study, which involves human participants, was approved by the Ethics Committee of Policlinico San Matteo in Pavia, Italy (P-20200055757, approval date 1 July 2020) and conducted following the declaration of Helsinki (1964) and its later amendment.

Informed Consent Statement: Informed consent was obtained from all subjects involved in the study. Written informed consent has been obtained from the patients to publish this paper. Written informed consent was provided by the participant's legal guardian/next of kin.

Data Availability Statement: Data are available upon reasonable request in Zenodo (10.5281/zenodo.7599703).

Acknowledgments: We thank the Italian Ministry of Health for supporting the study (Ricerca Corrente). We thank the patients and their families who took part in the study, and the Mondino Foundation Suicidality Research Group (Luca Capone, Antonio Piotti, Chiara Coci, Chiara Rogantini).

Conflicts of Interest: The authors declare no conflict of interest.

References

1. Ruch, D.A.; Bridge, J.A. Epidemiology of Suicide and Suicidal Behavior in Youth. In *Youth Suicide Prevention and Intervention: Best Practices and Policy Implications*; Ackerman, J.P., Horowitz, L.M., Eds.; SpringerBriefs in Psychology; Springer International Publishing: Cham, Switzerland, 2022; pp. 3–12, ISBN 978-3-031-06127-1.
2. Farooq, S.; Tunmore, J.; Ali, W.; Ayub, M. Suicide, self-harm and suicidal ideation during COVID-19: A systematic review. *Psychiatry Res.* **2021**, *306*, 114228. [CrossRef] [PubMed]
3. Cheng, H.; Wang, D.; Wang, L.; Zou, H.; Qu, Y. Global prevalence of self-harm during the COVID-19 pandemic: A systematic review and meta-analysis. *BMC Psychol.* **2023**, *11*, 149. [CrossRef] [PubMed]
4. Mensi, M.M.; Iacopelli, M.; Orlandi, M.; Capone, L.; Rogantini, C.; Vecchio, A.; Casini, E.; Borgatti, R. Psychiatric symptoms and emotional impact of the COVID-19 pandemic on Italian adolescents during the third lockdown: A cross-sectional cohort study. *Sci. Rep.* **2022**, *12*, 200901. [CrossRef] [PubMed]
5. Holman, M.S.; Williams, M.N. Suicide Risk and Protective Factors: A Network Approach. *Arch. Suicide Res.* **2022**, *26*, 137–154. [CrossRef] [PubMed]
6. Bilsen, J. Suicide and Youth: Risk Factors. *Front. Psychiatry* **2018**, *9*, 540. [CrossRef]
7. Turecki, G.; Brent, D.A.; Gunnell, D.; O'connor, R.C.; Oquendo, M.A.; Pirkis, J.; Stanley, B.H. Suicide and suicide risk. *Nat. Rev. Dis. Prim.* **2019**, *5*, 1–22. [CrossRef]
8. Miranda-Mendizabal, A.; Castellví, P.; Parés-Badell, O.; Alayo, I.; Almenara, J.; Alonso, I.; Blasco, M.J.; Cebrià, A.; Gabilondo, A.; Gili, M.; et al. Gender differences in suicidal behavior in adolescents and young adults: Systematic review and meta-analysis of longitudinal studies. *Int. J. Public Health* **2019**, *64*, 265–283. [CrossRef]
9. Callanan, V.J.; Davis, M.S. Gender differences in suicide methods. *Soc. Psychiatry Psychiatr. Epidemiol.* **2011**, *47*, 857–869. [CrossRef]
10. Tsirigotis, K.; Gruszczynski, W.; Tsirigotis, M. Gender differentiation in methods of suicide attempts. *Med. Sci. Monit.* **2011**, *17*, PH65–PH70. [CrossRef]
11. Su, Y.; Ye, C.; Xin, Q.; Si, T. Major depressive disorder with suicidal ideation or behavior in Chinese population: A scoping review of current evidence on disease assessment, burden, treatment and risk factors. *J. Affect. Disord.* **2023**, *340*, 732–742. [CrossRef]
12. Lundberg, J.; Cars, T.; Lampa, E.; Selling, K.E.; Leval, A.; Gannedahl, A.; Själin, M.; Björkholm, C.; Hellner, C. Determinants and Outcomes of Suicidal Behavior Among Patients with Major Depressive Disorder. *JAMA Psychiatry* **2023**. [CrossRef]
13. Yates, K.; Lång, U.; Cederlöf, M.; Boland, F.; Taylor, P.; Cannon, M.; McNicholas, F.; DeVylder, J.; Kelleher, I. Association of Psychotic Experiences with Subsequent Risk of Suicidal Ideation, Suicide Attempts, and Suicide Deaths. *JAMA Psychiatry* **2019**, *76*, 180–189. [CrossRef] [PubMed]
14. Coentre, R.; Fonseca, A.; Mendes, T.; Rebelo, A.; Fernandes, E.; Levy, P.; Góis, C.; Figueira, M.L. Suicidal behaviour after first-episode psychosis: Results from a 1-year longitudinal study in Portugal. *Ann. Gen. Psychiatry* **2021**, *20*, 35. [CrossRef] [PubMed]
15. Paris, J. Suicidality in Borderline Personality Disorder. *Medicina* **2019**, *55*, 223. [CrossRef] [PubMed]
16. Rizk, M.M.; Herzog, S.; Dugad, S.; Stanley, B. Suicide Risk and Addiction: The Impact of Alcohol and Opioid Use Disorders. *Curr. Addict. Rep.* **2021**, *8*, 194–207. [CrossRef]
17. Esang, M.; Ahmed, S. A Closer Look at Substance Use and Suicide. *Am. J. Psychiatry Resid. J.* **2018**, *13*, 6–8. [CrossRef]
18. Probert-Lindström, S.; Berge, J.; Westrin, Å.; Öjehagen, A.; Pavulans, K.S. Long-term risk factors for suicide in suicide attempters examined at a medical emergency in patient unit: Results from a 32-year follow-up study. *BMJ Open* **2020**, *10*, e038794. [CrossRef]
19. Madigan, A.; Daly, M. Socioeconomic status and depressive symptoms and suicidality: The role of subjective social status. *J. Affect. Disord.* **2023**, *326*, 36–43. [CrossRef]
20. Guldin, M.-B.; Li, J.; Pedersen, H.S.; Obel, C.; Agerbo, E.; Gissler, M.; Cnattingius, S.; Olsen, J.; Vestergaard, M. Incidence of Suicide Among Persons Who Had a Parent Who Died During Their Childhood: A Population-Based Cohort Study. *JAMA Psychiatry* **2015**, *72*, 1227–1234. [CrossRef]
21. Calderaro, M.; Baethge, C.; Bermpohl, F.; Gutwinski, S.; Schouler-Ocak, M.; Henssler, J. Offspring's risk for suicidal behaviour in relation to parental death by suicide: Systematic review and meta-analysis and a model for familial transmission of suicide. *Br. J. Psychiatry* **2022**, *220*, 121–129. [CrossRef]

22. Marraccini, M.E.; Griffin, D.; O'neill, J.C.; Martinez, R.R.; Chin, A.J.; Toole, E.N.; Grapin, S.L.; Naser, S.C. School Risk and Protective Factors of Suicide: A Cultural Model of Suicide Risk and Protective Factors in Schools. *Sch. Psychol. Rev.* **2022**, *51*, 266–289. [CrossRef] [PubMed]
23. Carballo, J.J.; Llorente, C.; Kehrmann, L.; Flamarique, I.; Zuddas, A.; Purper-Ouakil, D.; Hoekstra, P.J.; Coghill, D.; Schulze, U.M.E.; Dittmann, R.W.; et al. Psychosocial risk factors for suicidality in children and adolescents. *Eur. Child Adolesc. Psychiatry* **2020**, *29*, 759–776. [CrossRef] [PubMed]
24. de Jonge-Heesen, K.W.J.; Rasing, S.P.A.; Vermulst, A.A.; Engels, R.C.M.E.; Creemers, D.H.M. How to Cope with Perfectionism? Perfectionism as a Risk Factor for Suicidality and the Role of Cognitive Coping in Adolescents. *J. Ration. Cogn. Behav Ther.* **2020**, *39*, 201–216. [CrossRef]
25. Stefa-Missagli, S.; Unterrainer, H.-F.; Giupponi, G.; Holasek, S.-J.; Kapfhammer, H.-P.; Conca, A.; Sarlo, M.; Erbuto, D.; Rogante, E.; Moujaes-Droescher, H.; et al. Suicide and Personality Traits: A Multicenter Study of Austrian and Italian Psychiatric Patients and Students. *Suicide Life-Threat. Behav.* **2019**, *50*, 220–232. [CrossRef]
26. Che, Y.; Fang, D.; Cai, L.; Ran, H.; Chen, L.; Wang, S.; Peng, J.; Sun, H.; Liang, X.; Xiao, Y. Mediation of Resilience in the Association Between Personality Traits and Suicidal Ideation Among Chinese Adolescents. *Front. Psychol.* **2022**, *13*, 898318. [CrossRef]
27. Barber, J.G. Relative Misery and Youth Suicide. *Aust. N. Z. J. Psychiatry* **2001**, *35*, 49–57. [CrossRef]
28. O'Connor, R.C.; Rasmussen, S.; Miles, J.; Hawton, K. Self-harm in adolescents: Self-report survey in schools in Scotland. *Br. J. Psychiatry* **2009**, *194*, 68–72. [CrossRef]
29. Enns, M.W.; Cox, B.J.; Inayatulla, M. Personality Predictors of Outcome for Adolescents Hospitalized for Suicidal Ideation. *J. Am. Acad. Child Adolesc. Psychiatry* **2003**, *42*, 720–727. [CrossRef]
30. Chabrol, H.; Saint-Martin, C. Psychopathic Traits and Suicidal Ideation in High-School Students. *Arch. Suicide Res.* **2009**, *13*, 64–73. [CrossRef]
31. Csorba, J.; Dinya, E.; Ferencz, E.; Steiner, P.; Bertalan, G.; Zsadon, A. Novelty seeking: Difference between suicidal and non-suicidal Hungarian adolescent outpatients suffering from depression. *J. Affect. Disord.* **2010**, *120*, 217–220. [CrossRef]
32. Hirvikoski, T.; Jokinen, J. Personality traits in attempted and completed suicide. *Eur. Psychiatry* **2012**, *27*, 536–541. [CrossRef] [PubMed]
33. McCallum, S.M.; Batterham, P.J.; Christensen, H.; Werner-Seidler, A.; Nicolopoulos, A.; Newton, N.; Teesson, M.; Calear, A.L. Personality factors associated with suicidal ideation, plans and attempts in adolescents. *J. Affect. Disord.* **2022**, *310*, 135–141. [CrossRef] [PubMed]
34. Beautrais, A.L.; Joyce, P.R.; Mulder, R.T. Personality Traits and Cognitive Styles as Risk Factors for Serious Suicide Attempts among Young People. *Suicide Life-Threat. Behav.* **1999**, *29*, 37–47. [CrossRef]
35. Posner, K.; Brown, G.K.; Stanley, B.; Brent, D.A.; Yershova, K.V.; Oquendo, M.A.; Currier, G.W.; Melvin, G.A.; Greenhill, L.; Shen, S.; et al. The Columbia–Suicide Severity Rating Scale: Initial Validity and Internal Consistency Findings from Three Multisite Studies with Adolescents and Adults. *Am. J. Psychiatry* **2011**, *168*, 1266–1277. [CrossRef]
36. Wechsler, D. *Wechsler Intelligence Scale for Children*; Organizzazioni Speciali: Firenze, Italy, 2003.
37. Wechsler, D. *Wechsler Adult Intelligence Scale*; Organizzazioni Speciali: Firenze, Italy, 2008.
38. *Mensi Martina Maria 20210826_DEP_SUICID_MMM_01 2021*; Zenodo: Geneva, Switzerland, 2021.
39. Shaffer, D. A Children's Global Assessment Scale (CGAS). *Arch. Gen. Psychiatry* **1983**, *40*, 1228–1231. [CrossRef]
40. Butcher, J.N. Minnesota Multiphasic Personality Inventory. In *The Corsini Encyclopedia of Psychology*; Weiner, I.B., Craighead, W.E., Eds.; Wiley: Hoboken, NJ, USA, 2010; pp. 1–3. ISBN 978-0-470-17024-3.
41. Faul, F.; Erdfelder, E.; Buchner, A.; Lang, A.-G. Statistical power analyses using G*Power 3.1: Tests for correlation and regression analyses. *Behav. Res. Methods* **2009**, *41*, 1149–1160. [CrossRef]
42. Faul, F.; Erdfelder, E.; Lang, A.-G.; Buchner, A. G*Power 3: A flexible statistical power analysis program for the social, behavioral, and biomedical sciences. *Behav. Res. Methods* **2007**, *39*, 175–191. [CrossRef]
43. IBM Corp. *IBM SPSS Statistics for Windows, Version 27.0*; IBM Corp: Armonk, NY, USA, 2020.
44. Castaldo, L.; Serra, G.; Piga, S.; Reale, A.; Vicari, S. Suicidal behaviour and non-suicidal self-injury in children and adolescents seen at an Italian paediatric emergency department. *Ann. Dell'istituto Super. Sanita* **2020**, *56*, 303–314. [CrossRef]
45. Motillon-Toudic, C.; Walter, M.; Séguin, M.; Carrier, J.-D.; Berrouiguet, S.; Lemey, C. Social isolation and suicide risk: Literature review and perspectives. *Eur. Psychiatry* **2022**, *65*, e65. [CrossRef]
46. Pietropolli Charmet, G. Uccidersi: Il tentativo di suidicio in adolescenza. In *Collana di Psicologia Clinica e Psicoterapia*; R. Cortina: Milano, Italy, 2009; ISBN 978-88-6030-231-1.
47. Castellví, P.; Miranda-Mendizábal, A.; Alayo, I.; Parés-Badell, O.; Almenara, J.; Alonso, I.; Blasco, M.J.; Cebrià, A.; Gabilondo, A.; Gili, M.; et al. Assessing the Relationship Between School Failure and Suicidal Behavior in Adolescents and Young Adults: A Systematic Review and Meta-analysis of Longitudinal Studies. *Sch. Ment. Health Multidiscip. Res. Pract. J.* **2020**, *12*, 429–441. [CrossRef]
48. Pia, T.; Galynker, I.; Schuck, A.; Sinclair, C.; Ying, G.; Calati, R. Perfectionism and Prospective Near-Term Suicidal Thoughts and Behaviors: The Mediation of Fear of Humiliation and Suicide Crisis Syndrome. *Int. J. Environ. Res. Public Health* **2020**, *17*, 1424. [CrossRef] [PubMed]

49. Vera-Varela, C.; Mirón, P.C.M.; Barrigón, M.L.; Álvarez-García, R.; Portillo, P.; Chamorro, J.; Baca-García, E.; MEmind Study Group. Low Level of Agreement Between Self-Report and Clinical Assessment of Passive Suicidal Ideation. *Arch. Suicide Res.* **2022**, *26*, 1895–1910. [CrossRef] [PubMed]
50. Na, P.J.; Yaramala, S.R.; Kim, J.A.; Kim, H.; Goes, F.S.; Zandi, P.P.; Vande Voort, J.L.; Sutor, B.; Croarkin, P.; Bobo, W.V. The PHQ-9 Item 9 based screening for suicide risk: A validation study of the Patient Health Questionnaire (PHQ)−9 Item 9 with the Columbia Suicide Severity Rating Scale (C-SSRS). *J. Affect. Disord.* **2018**, *232*, 34–40. [CrossRef] [PubMed]
51. Zheng, Y.; Zhang, H.; Fan, Q. Discordance between family report and clinical assessment of suicide attempts: A prospective study from the emergency department. *Gen. Psychiatry* **2021**, *34*, e100576. [CrossRef] [PubMed]
52. Erdman, H.P.; Greist, J.H.; Gustafson, D.H.; Taves, J.E.; Klein, M.H. Suicide risk prediction by computer interview: A prospective study. *J. Clin. Psychiatry* **1987**, *48*, 464–467. [PubMed]
53. Greist, J.H.; Laughren, T.P.; Gustafson, D.H.; Stauss, F.F.; Rowse, G.L.; Chiles, J.A. A Computer Interview for Suicide-Risk Prediction. *Am. J. Psychiatry* **1973**, *130*, 1327–1332. [CrossRef]
54. Sendbuehler, J.M.; Kincel, R.L.; Nemeth, G.; Oertel, J. Dimension of Seriousness in Attempted Suicide: Significance of the MF Scale in Suicidal MMPI Profiles. *Psychol. Rep.* **1979**, *44*, 343–361. [CrossRef]
55. Faruki, F.; Patel, A.; Jaka, S.; Kaur, M.; Sejdiu, A.; Bajwa, A.; Patel, R.S. Gender Dysphoria in Pediatric and Transitional-Aged Youth Hospitalized for Suicidal Behaviors: A Cross-National Inpatient Study. *Prim. Care Companion CNS Disord.* **2023**, *25*, 46284. [CrossRef]
56. Mitchell, H.K.; Keim, G.; Apple, D.E.; Lett, E.; Zisk, A.; Dowshen, N.L.; Yehya, N. Prevalence of gender dysphoria and suicidality and self-harm in a national database of paediatric inpatients in the USA: A population-based, serial cross-sectional study. *Lancet Child Adolesc. Health* **2022**, *6*, 876–884. [CrossRef]
57. Gipson, P.Y.; Agarwala, P.; Opperman, K.J.; Horwitz, A.; King, C.A. Columbia-Suicide Severity Rating Scale: Predictive Validity with Adolescent Psychiatric Emergency Patients. *Pediatr. Emerg. Care* **2015**, *31*, 88–94. [CrossRef]
58. Mensi, M.M.; Capone, L.; Rogantini, C.; Orlandi, M.; Ballante, E.; Borgatti, R. COVID-19-related psychiatric impact on Italian adolescent population: A cross-sectional cohort study. *J. Community Psychol.* **2021**, *49*, 1457–1469. [CrossRef] [PubMed]
59. Orbach, I.; Milstein, I.; Har-Even, D.; Apter, A.; Tiano, S.; Elizur, A. A Multi-Attitude Suicide Tendency Scale for adolescents. *Psychol. Assess. J. Consult. Clin. Psychol.* **1991**, *3*, 398–404. [CrossRef]
60. Benchimol, E.I.; Smeeth, L.; Guttmann, A.; Harron, K.; Moher, D.; Petersen, I.; Sørensen, H.T.; von Elm, E.; Langan, S.M.; The RECORD Working Committee. The REporting of studies Conducted using Observational Routinely-collected health Data (RECORD) Statement. *PLoS Med.* **2015**, *12*, e1001885. [CrossRef] [PubMed]

children

MDPI

Article

Psychiatric Hospitalization for Psychopathological Disorders and Self-Injurious Behaviors in Italian Children and Adolescents during COVID-19

Maria Pontillo [1,*], Deny Menghini [1], Roberto Averna [1], Milena Labonia [1], Giulia Lazzaro [1], Maria Cristina Tata [1] and Stefano Vicari [1,2]

1 Child & Adolescent Neuropsychiatry Unit, Bambino Gesù Children's Hospital, IRCCS, 00146 Rome, Italy; deny.menghini@opbg.net (D.M.); roberto.averna@opbg.net (R.A.); milena.labonia@opbg.net (M.L.); giulia.lazzaro@opbg.net (G.L.); mariacristina.tata@opbg.net (M.C.T.); stefano.vicari@opbg.net (S.V.)
2 Department of Life Sciences and Public Health, Università Cattolica del Sacro Cuore, 00168 Rome, Italy
* Correspondence: maria.pontillo@opbg.net

Abstract: The evidence shows that the COVID-19 pandemic dramatically increased the number of urgent psychiatric consultations for children and adolescents in hospital emergency departments (EDs). However, what needs to be further investigated are the characteristics of psychiatric hospitalization in children and adolescents admitted to the Child and Adolescent Neuropsychiatry Unit wards in EDs. Specifically, this retrospective study aimed to examine changes in (i) the number of inpatients and (ii) the distribution of psychopathological disorders and self-injurious behaviors in our Child and Adolescent Neuropsychiatry Unit ward during the COVID-19 lockdown in Italy (March–June 2020; October 2020–January 2021) compared with the same months of previous years. We found a significantly lower number of inpatients during the first four quarantine months than the first four reference months and a higher number of inpatients during the second four quarantine months than the second four reference months. Additionally, we found an increased frequency of mood disorders, non-suicidal self-injurious behavior, and suicidal ideation during the COVID-19 lockdown compared to the reference periods. Our findings underline the need to develop psychological healthcare services for future emergency periods in order to identify and treat psychological distress in children and adolescents early, reducing the risk of psychiatric hospitalization.

Keywords: COVID-19 pandemic; children; adolescents; psychiatric hospitalization; mood disorders; non-suicidal self-injurious behaviors; suicidal ideation

Citation: Pontillo, M.; Menghini, D.; Averna, R.; Labonia, M.; Lazzaro, G.; Tata, M.C.; Vicari, S. Psychiatric Hospitalization for Psychopathological Disorders and Self-Injurious Behaviors in Italian Children and Adolescents during COVID-19. *Children* **2023**, *10*, 1846. https://doi.org/10.3390/children10121846

Academic Editor: Brian Littlechild

Received: 11 October 2023
Revised: 10 November 2023
Accepted: 23 November 2023
Published: 24 November 2023

1. Introduction

In March 2020, the novel coronavirus (i.e., COVID-19) pandemic forced the general population to engage with unprecedented social distancing and isolation measures in an effort to contain the infection. Worldwide, governments adopted strict lockdown measures to control the spread of the virus. Therefore, everyday habits and lifestyles were dramatically altered across familial, social, and professional spheres. This had a profound impact on the functioning of all individuals, regardless of their age, sex, and socio-economic status.

Research has shown that viral outbreaks and their associated quarantine measures tend to decrease psychosocial wellbeing. During the SARS CoV1 epidemic, for example, major anxiety and depression symptoms were found to be associated with a major risk of developing post-traumatic stress disorder and abusive conduct in young adults [1–3]. A number of studies [4–7] have reported a widespread presence of anxiety symptoms, mood deflection, and sleep disturbance in the general population during the COVID-19 pandemic. They have also reported significant psychological effects in children and adolescents [8–10], manifesting differently according to age. Specifically, in preschool children (i.e., aged

3–6 years), Jiao et al. [11] found an increase in irritability, inattention, and disruptive behaviors during the COVID-19 pandemic, associated with a high incidence of sleep disturbance, agitation, and separation anxiety due to the fear that family members could contract the virus. In children and adolescents (i.e., aged 6–18 years), Uccella et al. [12] observed an association between isolation, social distancing due to COVID-19, and anxiety, fear, and uncertainty about the future.

The COVID-19 pandemic and its associated restrictions may have particularly affected adolescents and preadolescents, who would have already been experiencing a vulnerable transitional period characterized by the onset of puberty and significant neurobiological, social, and cognitive changes [13,14]. Specifically, adolescents may have experienced aggravated chronic and acute stress during the pandemic, in the form of worrying about family members, unexpected mourning, anxiety around the sudden closure of school and home confinement, stress caused by an increased use of the internet and social media, and worry about the economic future of their family. In this vein, Vindegaard and Benros (5) suggested that the interruption of social relations during periods of school closure represented a main stressor for adolescents during the pandemic.

Literature narrative review from Guessoum et al. [15] on adolescent mental health during the COVID-19 pandemic showed an increase in psychopathological disorders, including depressive disorders, anxiety disorders, post-traumatic stress disorder, and grief-related symptoms. In a retrospective cohort study, Ougrin et al. [16] found that the proportion of young people presenting with self-harm increased from 50% in 2019 to 57% in 2020; moreover, the proportion of young people with emotional disorders increased from 58% to 66% over the same period.

In a recent meta-analysis, Panchal et al. (2023) examined in detail this significant existing literature on the effects of quarantine measures on the mental health of children and adolescents.

The results highlighted that anxiety and depression are the most frequently reported symptoms by children and adolescents during quarantine periods. Specifically, the prevalence of anxiety symptoms varied between 1.8% and 49.5% in the studies, with 13.4% of the examined children reporting severe anxiety. The prevalence of depressive symptoms ranged from 2.2% to 63.8% in the studies, and 27% of the children and adolescents included in the studies reported severe depression. Additionally, irritability was frequently reported among children and adolescents during quarantine periods, with a prevalence ranging from 16.7% to 73.2% in the various studies examined. Panchal et al. [17], along with another review by Theberath et al. [18], concluded by emphasizing the need for in-depth studies on the effects of lockdown on the mental health of children and adolescents. These studies should facilitate the development of appropriate guidelines for managing emergency periods like the one that has passed. In fact, most of the studies cited here primarily focus on the prevalence of psychiatric symptoms in children and adolescents during COVID-19 and associated quarantine measures without delving into the different levels of care (e.g., urgent psychiatric consultation and psychiatric hospitalization) required for these symptoms during that time. We believe that this is necessary to fully understand this impact and prepare clinical services for future periods in which significant reductions in social, school, and leisure activities may once again be enforced.

This is particularly relevant to the context of Italy, which was the first European country to adopt strict measures involving social distancing, prohibitions, and restrictions during the COVID-19 pandemic

In February 2020, the Italian government declared a state of emergency over the COVID-19 pandemic. The first phase of the government's COVID-19 response (9 March to 3 May 2020) included a general school closure, freedom of movement restrictions (enforceable by law), and the cancellation of all non-essential trips (e.g., school trips abroad). Consequently, all activities were paused, resulting in a significant reduction in child and adolescent extracurricular activities, including sports practices, music lessons, and theater classes. During the second phase of the COVID-19 response (4 May to 14 June 2020), containment measures were eased, resulting in the legal removal of social isolation and regional movement limitations and the resumption of commercial activities. However, schools remained closed, and only exams were attended in person, with restrictions. Finally, in the third stage of the COVID-19 response (15 June to 7 October 2020), all leisure activities were resumed with restrictions.

The second wave of COVID-19 diffusion was observable in Italy from 8 October 2020. At this time, the Italian government once again applied strong limitations, closing schools and non-essential services. Furthermore, beginning on 6 November 2020, the Italian government, following European Community directions, applied regional containment measures, based on the regional contagious index (i.e., the Rt index). At that time, school attendance was limited to a percentage capacity (e.g., 50% of all students attending at a time). Finally, from 7 January 2021, regional restrictions continued, enforceable via colored bands for each region.

Previous retrospective studies conducted by our research group have focused on the increase in urgent psychiatric consultations for children and adolescents during the COVID-19 pandemic in Italy. In particular, Serra et al. [19] demonstrated that urgent psychiatric consultations for any reason and for suicide attempts significantly reached their peak during the second quarantine period in Italy. However, this study examines data from our emergency department and does not take into account any changes in the psychiatric hospitalization of children and adolescents during the quarantine periods. We believe that this should be investigated in light of the prolonged period of home confinement, school closures, and social restrictions experienced by Italian children and adolescents. Based on these considerations, our study is the first initial examination of the characteristics of psychiatric hospitalization in Italian children and adolescents during COVID-19. Specifically, in the current retrospective study, we examined changes in the number of admissions to the Child and Adolescent Neuropsychiatry Unit ward of Bambino Gesù Children's Hospital and the presentation of psychopathological disorders (i.e., psychosis, mood disorder, post-traumatic stress disorder, and behavior disorder) and self-injurious behaviors in the admitted patients during the COVID-19 quarantine periods in Italy, in comparison to reference periods. Specifically, admissions during the months in which quarantine was imposed in Italy (i.e., March to June 2020; October 2020 to January 2021) were compared with admissions in the 12 months prior (i.e., March to June 2019; October 2019 to January 2020).

2. Materials and Methods

2.1. Participants

The present retrospective study examined admissions to the Child and Adolescent Neuropsychiatry Unit ward of Bambino Gesù Children's Hospital during COVID-19 pandemic based on a retrospective chart review. Bambino Gesù Children's Hospital is the largest pediatric hospital in Europe and a 607-bed tertiary care academic hospital in Italy (Rome). The hospital is widely recognized as a referral center for all pediatric specialties at the national and international levels. The Child and Adolescent Neuropsychiatry Unit emergency ward admits children and adolescents (i.e., aged 5–17.9 years) who have been referred from the emergency department at a maximum rate of eight beds per day.

For our analysis, we included data on patients (a) aged 6–17.9 years and (b) who have been referred from the emergency department only once. Patients who presented with repeated evaluations and/or the revolving door phenomenon were not considered for the current study. The current study examined admissions through a retrospective analysis of medical records during two consecutive 4-month periods when a national quarantine was enforced in response to the COVID-19 pandemic in Italy: from March to June 2020 and October 2020 to January 2021. Admissions during the corresponding periods from 12 months prior (i.e., March to June 2019 and October 2019 to January 2020) were also taken into account, enabling comparisons regarding the number of admissions as well as the distribution of psychopathological disorders and self-injurious behaviors.

In our study, we considered patients with psychopathological disorders diagnosed according to DSM-5 criteria (i.e., psychosis, mood disorders, post-traumatic stress disorder, and behavior disorders) and self-injurious behaviors defined as "any act of self-poisoning or self-injury, irrespective of the underlying intent," according to the UK National Institute for Health and Care Excellence [20] clinical guidelines. Thus, both non-suicidal self-injury and suicide attempts were included. Suicidal ideation was categorized, according to the DSM-5 [21], as "thoughts about self-harm, with deliberate consideration or planning of possible techniques of causing one's own death".

The study was approved by the Ethics Committee of the Children Hospital Bambino Gesù. All participants and their parents/legal guardians provided written informed assent and consent.

2.1.1. Clinical Setting

In our Child and Adolescent Neuropsychiatry Unit ward, psychiatric hospitalization is conducted as follows: continuous care for admitted children and adolescents is provided by nurses and health and social workers, as parental presence within the ward is prohibited. Parents have scheduled visits with their children each afternoon, lasting approximately 2 h. Limiting parental presence serves the purpose of preventing intra-family aggression or crises and facilitating an assessment of the patient free from parental influence on the reported information. Nevertheless, parents play an integral role during their children's hospitalization. Psychologists and neuropsychiatrists engage with parents for diagnostic evaluations and updates on their children's clinical conditions. Psychoeducational interventions are designed to alleviate parental stress related to the diagnosis and hospitalization of their children and adolescents. Moreover, neuropsychiatrists and psychologists engage with children and adolescents admitted to daily group therapy sessions, psychological assessment tests, and individual psychological interventions. Regular team meetings are scheduled between nurses, neuropsychiatrists, and psychologists to coordinate tasks and address critical issues. Surveillance cameras are installed in all rooms to monitor any instances of self-harm or suicidal attempts; a central monitor is positioned in the nursing room. The ward also includes a common area where children participate in various activities together, such as shared meals, educational sessions conducted by teachers according to their educational level, drama classes, music listening, and movie screenings.

The primary objective of hospitalization is to address the acute phase of the disorder and to facilitate the transition to mental health departments within the local community. Certain inpatients may be transferred to specialized psychiatric facilities catering to children and adolescents experiencing severe emotional disorders and/or behavioral issues that cannot be adequately managed in outpatient settings.

2.1.2. Clinical Assessment

All the patients included in this study were assessed by expert neuropsychiatrists and developmental psychologists according to DSM-5 criteria [21], based on developmental history and an extensive clinical examination. All the patients considered were evaluated via the same psychopathological assessment protocol:

- Psychopathological disorders were assessed using the Schedule for Affective Disorders and Schizophrenia for School-Aged Children [22], a semi-structured interview administered to obtain a psychopathological diagnosis according to the DSM-5 criteria. The K-SADS-PL DSM-5, as proposed in the instrument manual by Kaufman et al. [22], provides a source of information for the child/adolescent as well as the parents.
- Suicidal ideation and behavior were assessed with the Columbia–Suicide Severity Rating Scale (C-SSRS). Suicidal ideation is defined by a score of 3 or above.
- Non-suicidal self-injury (NSSI) was evaluated with an assessment of the DSM-5 categorical diagnostic criteria for non-suicidal self-injury (NSSI), that is, NSSI on at least 5 days within the past year, suicidal ideation absent or a low score (a score

below 3 at the C-SSRS), and no previous suicide attempts. All these diagnostic tools were administered by trained child neuropsychiatrists and psychologists.

2.2. Statistical Analyses

The number of admissions was first considered for the statistical analyses and refers to the number of patients who have only accessed the Child and Adolescent Neuropsychiatry Unit ward of the Bambino Gesù Children's Hospital once. Patients presenting revolving-door and repeated evaluations during the analyzed period were excluded. Chi-square (χ^2) contingency tables were used to explore potential differences in the number of admissions between the first quarantine period (March to June 2020) and the first reference period (March to June 2019), as well as for the second quarantine period (October 2020 to January 2021) compared to the second reference period (October 2019 to January 2020). To verify which cells contributed to potential significant differences, calculations of residuals proposed by Sharpe et al. [23] were conducted.

Moreover, the number of psychopathological disorders was considered for the statistical analyses, and the number of patients with psychopathological disorders as a primary diagnosis, i.e., psychosis, mood disorders, post-traumatic disorders, and behavior disorders, was calculated among the total of admissions. χ^2 contingency tables were used, and odds ratios (ORs) with 95% confidence intervals (95% CI) were calculated to measure the associations between diagnoses (i.e., psychosis, mood disorder, post-traumatic stress disorder, and behavior disorder) and the quarantine period and the reference periods.

The same analyses were run to explore the associations between self-injurious behaviors (i.e., suicidal ideation, non-suicidal self-injurious behaviors, and suicide attempts) and the quarantine versus the reference periods.

A p-value of less than 0.05 was considered statistically significant. χ^2 analyses and OR were computed using R Studio (R Studio, Boston, MA, USA), with particular reliance on the Epitools package.

3. Results

3.1. Number of Admissions

The total number of admissions among these periods was 377, with an average of 6 days of hospitalization and ages ranging from 6 to 18 years old. The number of admissions to the Child and Adolescent Neuropsychiatry Unit ward differed between the quarantine and the reference periods ($\chi 21 = 5.43$, $p = 0.0197$). Specifically, z-tests were run to compare cells [23], with the critical value of 2.33 ($\sqrt{\chi 21} = \sqrt{5.43}$). There were significantly fewer admissions during the first quarantine period than the first reference period (77 vs. 91; z-test = -2.34) and significantly more admissions during the second quarantine period than the second reference period (121 vs. 88; z-test = 2.34).

3.2. Distribution of Psychopathological Disorders

For a more comprehensive overview of the psychopathological distribution trend, we analyzed the distributions of psychopathological disorders, merging those from the first quarantine period (March to June 2020) and the second quarantine period (October 2020 to January 2021), as well as the first reference period (March to June 2019) and the second reference period (October 2019 to January 2020). See Table 1.

Overall, the results showed a significant association between mood disorders and the quarantine versus the reference periods, with significantly more diagnosed mood disorders among admitted patients during the quarantine periods (OR = 1.54, 95% CI [1.01, 2.37], $p = 0.04$). No further associations emerged between diagnoses and the quarantine period and the reference periods (psychosis: OR = 0.68, 95% CI [0.41, 1.12], $p = 0.13$; post-traumatic stress disorder: OR = 1.01, 95% CI [0.50, 2.04], $p = 0.97$; behavioral disorders: OR = 0.50, 95% CI [0.19, 1.21], $p = 0.12$).

Table 1. Distribution of psychopathological diagnoses between the quarantine and the reference periods.

	Distribution among Admitted Patients	
Diagnosis	**Quarantine Periods**	**Reference Periods**
	# (%)	**# (%)**
Psychosis	34 (17.2)	42 (23.5)
Mood disorders	137 (69.2)	106 (59.2)
Post-traumatic stress disorder	19 (9.6)	17 (9.5)
Behavior disorders	8 (4.0)	14 (7.8)

The distributions of psychopathological disorders (i.e., psychosis, mood disorder, post-traumatic stress disorder, and behavior disorder) between the first quarantine period (March to June 2020) and the first reference period (March to June 2019), as well as for the second quarantine period (October 2020 to January 2021) and the second reference period (October 2019 to January 2020), are presented in the Supplementary Materials (Table S1).

3.3. Distribution of Self-Injurious Behaviors

For a more comprehensive overview of the self-injurious behaviors trend, we analyzed the distributions of self-injurious behaviors (i.e., suicidal ideation, non-suicidal self-injurious behaviors, and suicide attempts), merging those from the first quarantine period (March to June 2020) and the second quarantine period (October 2020 to January 2021), as well as the first reference period (March to June 2019) and the second reference period (October 2019 to January 2020). See Table 2.

Table 2. Distribution of self-injurious behaviors between the quarantine and the reference periods.

	Distribution among Admitted Patients	
Self-Injurious Behaviors	**Quarantine Periods**	**Reference Periods**
	# (%)	**# (%)**
Suicidal ideation	80 (40.4)	53 (29.6)
Non-suicidal self-injurious behaviors	116 (58.6)	77 (43.2)
Suicide attempts	34 (17.2)	34 (18.9)

The results showed a significant association between suicidal ideation and the quarantine and the reference periods, with significantly more suicidal ideation among admitted patients during the quarantine periods (OR = 1.61, 95% CI [1.05, 2.48], p = 0.03). Furthermore, a significant association emerged between non-suicidal self-injurious behaviors and the quarantine and the reference periods, revealing a greater frequency of non-suicidal self-injurious behaviors among admitted patients during the quarantine periods (OR = 1.87, 95% CI [1.24, 2.82], p = 0.003). No significant association was found with reference to suicide attempts (OR = 0.88, 95% CI [0.52, 1.5], p = 0.65).

The distributions of self-injurious behaviors (i.e., suicidal ideation, non-suicidal self-injurious behaviors, and suicide attempts) between the first quarantine period (March to June 2020) and the first reference period (March to June 2019), as well as for the second quarantine period (October 2020 to January 2021) and the second reference period (October 2019 to January 2020), are presented in Supplementary Materials (Table S2).

4. Discussion

The present retrospective study examined changes in the number of direct admissions to the Child and Adolescent Neuropsychiatry Unit emergency ward of Bambino Gesù Children's Hospital and the distribution of psychiatric disorders and self-injurious behaviors among admitted patients during the COVID-19 quarantine in Italy, in comparison to a reference period.

The results showed a decrease in the number of direct admissions during the first quarantine period.

Specifically, the number of admissions during the first quarantine period (77 patients, from March to June 2020) was lower than that of the first reference period (91 patients, from March to June 2019). This finding aligns with the results of a retrospective international cohort study [16] of children and adolescent inpatients in hospital emergency departments in European countries (i.e., England, Scotland, Ireland, Hungary, Turkey, etc.).

There are two potential explanations for the lower number of admissions during the first quarantine period. The first explanation is that families may have avoided emergency departments for fear of becoming infected. Our hypothesis is consistent with the results of previous studies [24,25], which have underlined that the worries of young people and their parents about contracting and spreading COVID-19 may have severely reduced young people's psychological wellbeing and the use of mental health services. The second explanation, as noted by Ougrin et al. [16], is that the lower number of children and adolescent admissions to psychiatry wards may be explained by the quarantine measures at the time. In Italy, many children and adolescents stopped attending school in March 2020. This is likely to have reduced their experience of academic pressure and minimized their number of distress factors. For example, many children and adolescents may have had fewer or no face-to-face relationships, which the literature [23] associates with an increased risk of self-injurious behaviors and psychopathological disorders, as well as bullying and peer pressure to abuse alcohol and drugs. In addition, many children and adolescents may have increased their psychological wellbeing during the first quarantine period as a result of spending more time with family.

Accordingly, it may be that some children and adolescents who would have otherwise accessed the hospital in crisis were able to access alternative coping strategies linked to different habits (i.e., staying at home) during the lockdown. However, staying at home may have also generated long-term negative effects on the mental health of children and adolescents. In line with this, the present study found that the number of direct admissions increased significantly during the second quarantine period compared to the second reference period (i.e., 121 patients from October 2020 to January 2021 vs. 88 patients from October 2019 to January 2020). It is possible that, in the second quarantine period, the persistent isolation, social distancing, and homeschooling may have determined anxiety, distress, fear, and a low mood, while reducing the ability to perform coping strategies and increasing the need for psychiatric services. In addition, differently from the first quarantine period, in the second quarantine period, parents went back to work, and they were thus less able to provide physical and emotional support to their children, who nevertheless continued to stay at home.

Considering the distribution of psychopathological disorders among the child and adolescent admissions during the quarantine periods, mood disorders were the most common. Additionally, these were more frequent during the quarantine periods than the reference periods (69.2% vs. 59.2%, respectively). These results are compatible with the literature [25–28], which indicates that the persistence of social restrictions and isolation during the COVID-19 pandemic was a risk factor for psychopathological disorders (including mood disorders) among children and adolescents. Oliva et al. [29], in a prospective cross-sectional study involving 9688 pre-adolescents and adolescents, showed that the lifestyle changes generated by enforced and prolonged social isolation were associated with an increased prevalence of mental disorders, including mood disorders. In particular, the reduction in physical activity, the increased screen exposure (including for the purposes of homeschooling), and the interruption of face-to-face relationships with peers (which fostered a dependence on social media that contributed to the excessive screen exposure) were identified as risk factors for the psychological wellbeing of children and adolescents during COVID-19. Additionally, changes in sleeping patterns (e.g., irregular wake-up times and shorter or longer periods of sleep) were also acknowledged as a risk factor [30].

Concerning self-injurious behaviors among the admitted patients during the quarantine periods, an increased frequency of non-suicidal self-injurious behaviors was observed compared to the reference periods (i.e., 58.6% during the quarantine periods vs. 43.2% during the reference periods). This finding aligns with the results of previous studies of adolescents during COVID-19 [16,31–33]. As proposed by Plener [30], non-suicidal self-injurious behaviors are often applied as an emotion regulation strategy to temporarily decrease or eliminate negative emotions. Therefore, negative emotions activated by lockdown pressure may have represented a significant risk factor for non-suicidal self-injurious behaviors during the quarantine. Additionally, during the quarantine periods, social support, which has been identified as a strong protective factor against non-suicidal self-injurious behaviors [34,35], may have only been accessible to some children and adolescents through social media. These factors, alongside limited access to face-to-face professional support, may have amplified children and adolescents' need to regulate negative emotions, and thus increased their likelihood of practicing non-suicidal self-injurious behaviors. In line with the literature [36–38], the present findings could be helpful in informing the development of clinical and educational intervention programs for non-suicidal self-injurious behaviors during quarantine periods.

Moreover, Asarnow et al. [39] described non-suicidal self-injurious behaviors as a significant predictor of suicidal ideation and suicide attempts in adolescence. Consistent with this, the present study found increased suicidal ideation amongst children and adolescent admissions during the quarantine periods (i.e., 40.4% during the quarantine periods vs. 29.6% during the reference periods). However, no increase in suicidal attempts during the quarantine periods was observed (i.e., 17.2% during the quarantine periods vs. 18.9% during the reference periods). As a potential explanation for this, home confinement may have facilitated child–parent communication and intensified parental surveillance, thereby minimizing suicide attempts.

At the time of writing, the present study represented the first to assess the characteristics of psychiatric hospitalization, such as the number of inpatients admitted, psychopathological disorders, and self-injurious behavior, in children and adolescents during the two COVID-19 quarantine periods in Italy. Regarding the limitations of our study, this research did not delve into the relationship between psychopathological disorders or self-injurious behaviors detected during hospitalization and any risk factors. These risk factors could be clinical (e.g., psychiatric family history or prior psychiatric diagnoses), individual (e.g., temperament traits or personality characteristics), or environmental (emotional or physical abuse within the family and intra-family conflicts). These factors, along with lifestyle changes associated with quarantine measures (e.g., changes in sleep patterns due to home confinement, homeschooling, and decreased face-to-face social interaction), should be considered in future research. Furthermore, future studies could investigate the long-term outcomes for children and adolescents hospitalized during the COVID-19 pandemic. It would be interesting to examine whether there is a chronicity of the mood disorders identified and the clinical progression of the self-injurious behaviors. This would allow for further reflections on interventions, both preventive and therapeutic, to be implemented in emergency situations.

5. Conclusions

The present findings offer valuable insights into the impact of the COVID-19 pandemic and quarantine measures on psychiatric hospitalization among Italian children and adolescents. Our results highlight changes in admission patterns during the COVID-19 pandemic and suggest a critical examination of the psychological needs of young individuals during emergency situations. Children and adolescents are particularly susceptible to the psychological effects of the COVID-19 pandemic due to their developing ability to regulate emotions and cope with stressful life events. Consequently, they may exhibit negative psychological outcomes requiring psychiatric hospitalization. In our sample of Italian children and adolescents, we found an increased frequency of mood disorders, non-suicidal self-injurious behavior, and suicidal ideation during the COVID-19 lockdown compared to reference periods. Although with caution, we hypothesize that the rapid spread of COVID-19 in Italy with numerous victims and the consequent prolonged period of home isolation with the destruction of daily routines, the closure of schools, and the absence of in-person relationships have represented factors precipitating the onset of mood disorders in probably already vulnerable young people. Moreover, despite the implementation of home-schooling with the possibility of meeting teachers online and the presence of parents at home, it should be considered that the negative emotional atmosphere and fear of death present both in the school and family contexts have hindered the long-term adaptation of young people to the pandemic and the quarantine measures necessary for its containment.

In times of emergency, supportive therapeutic programs can prevent the distress of hospitalization and separation from home and family. Drawing from the cooperative model proposed by Brofenbrenner [40] and Zhou [41], we propose psychological interventions involving three cooperative systems: the social system, the school system, and the family system. The social system (government and social organizations) could implement psychological screening programs (via social media, online chat, or telephone) to identify and address early signs of anxiety, depression, and emotional dysregulation among children and adolescents experiencing a pandemic and associated quarantine measures.

Simultaneously, involving parents and teachers in these psychological support programs is essential to alleviate their emotional distress, enhance their coping strategies, and help children and adolescents. These support programs should be accessible during both the acute and post-pandemic stages. In the acute phase, they can shield children and adolescents from psychological and behavioral crises that might lead to psychiatric hospitalization. In the post-pandemic phase, these programs serve to protect them from enduring long-term psychological difficulties that could potentially progress to diagnosable psychiatric disorders.

Supplementary Materials: The following supporting information can be downloaded at https://www.mdpi.com/article/10.3390/children10121846/s1. Table S1. Distribution of psychopathological diagnoses between the first quarantine period, first reference period, second quarantine period, and the second reference period. Table S2. Distribution of self-injurious behaviors between the first quarantine period, first reference period, second quarantine period, and the second reference period.

Author Contributions: Conceptualization, M.P. and D.M.; Methodology, M.P. and R.A.; Formal Analysis, D.M. and G.L.; Writing—Original Draft Preparation, M.P., D.M., R.A., M.L. and G.L.; Writing—Review and Editing, M.P., M.C.T. and S.V.; Supervision, D.M. and S.V. All authors have read and agreed to the published version of the manuscript.

Funding: This work was supported by the Italian Ministry of Health with Current Research 194 funds.

Institutional Review Board Statement: The study was conducted in accordance with the Declaration of Helsinki and in accordance with Italian legal and ethical requirements for clinical data. IRB approval was obtained for the data reporting in the present report (Ethics Committee of Bambino Gesù Children's Hospital practice number 3161/2023). Ethical approval date: 27 June 2023.

Informed Consent Statement: Informed consent was obtained from all subjects involved in the study.

Data Availability Statement: The raw data that underpins the conclusions of this article will be provided by the corresponding author upon reasonable request. The data is not publicly available due to the nature of the information and the regulations governing the retrospective study mode, which involves the review of medical records.

Conflicts of Interest: The authors declare no conflict of interest.

References

1. Hawryluck, L.; Gold, W.L.; Robinson, S.; Pogorski, S.; Galea, S.; Styra, R. SARS Control and Psychological Effects of Quarantine, Toronto, Canada. *Emerg. Infect. Dis.* **2004**, *10*, 1206–1212. [CrossRef] [PubMed]
2. Reynolds, D.L.; Garay, J.R.; Deamond, S.L.; Moran, M.K.; Gold, W.; Styra, R. Understanding, compliance and psychological impact of the SARS quarantine experience. *Epidemiol. Infect.* **2008**, *136*, 997–1007. [CrossRef] [PubMed]
3. Brooks, S.K.; Smith, L.E.; Webster, R.K.; Weston, D.; Woodland, L.; Hall, I.; Rubin, G.J. The impact of unplanned school closure on children's social contact: Rapid evidence review. *Eurosurveillance. Euro Surveill.* **2020**, *25*, 2000188. Available online: https://www.eurosurveillance.org/content/10.2807/1560-7917.ES.2020.25.13.2000188 (accessed on 2 April 2020). [CrossRef] [PubMed]
4. Torales, J.; O'Higgins, M.; Castaldelli-Maia, J.M.; Ventriglio, A. The outbreak of COVID-19 coronavirus and its impact on global mental health. *Int. J. Soc. Psychiatry* **2020**, *66*, 317–320. [CrossRef] [PubMed]
5. Vindegaard, N.; Benros, M.E. COVID-19 pandemic and mental health consequences: Systematic review of the current evidence. *Brain Behav. Immun.* **2020**, *89*, 531–542. [CrossRef] [PubMed]
6. Racine, N.; McArthur, B.A.; Cooke, J.E.; Eirich, R.; Zhu, J.; Madigan, S. Global Prevalence of Depressive and Anxiety Symptoms in Children and Adolescents During COVID-19: A Meta-analysis. *JAMA Pediatr.* **2021**, *175*, 1142. [CrossRef]
7. Wolf, K.; Schmitz, J. Scoping review: Longitudinal effects of the COVID-19 pandemic on child and adolescent mental health. *Eur. Child. Adolesc. Psychiatry* **2023**, *17*, 104. [CrossRef]
8. Shen, K.; Yang, Y.; Wang, T.; Zhao, D.; Jiang, Y.; Jin, R.; Zheng, Y.; Xu, B.; Xie, Z.; Lin, L.; et al. Diagnosis, treatment, and prevention of 2019 novel coronavirus infection in children: Experts' consensus statement. *World J. Pediatr.* **2020**, *16*, 223–231. [CrossRef]
9. Meherali, S.; Punjani, N.; Louie-Poon, S.; Abdul Rahim, K.; Das, J.K.; Salam, R.A.; Lassi, Z.S. Mental Health of Children and Adolescents Amidst COVID-19 and Past Pandemics: A Rapid Systematic Review. *Int. J. Environ. Res. Public Health* **2021**, *18*, 3432. [CrossRef]
10. Samji, H.; Wu, J.; Ladak, A.; Vossen, C.; Stewart, E.; Dove, N.; Long, D.; Snell, G. Review: Mental health impacts of the COVID-19 pandemic on children and youth—A systematic review. *Child. Adoles Ment. Health* **2022**, *27*, 173–189. [CrossRef]
11. Jiao, W.Y.; Wang, L.N.; Liu, J.; Fang, S.F.; Jiao, F.Y.; Pettoello-Mantovani, M.; Somekh, E. Behavioral and Emotional Disorders in Children during the COVID-19 Epidemic. *J. Pediatr.* **2020**, *21*, 264–266.e1. [CrossRef] [PubMed]
12. Uccella, S.; De Grandis, E.; De Carli, F.; D'Apruzzo, M.; Siri, L.; Preiti, D.; Di Profio, S.; Rebora, S.; Cimellaro, P.; Biolcati Rinaldi, A.; et al. Impact of the COVID-19 Outbreak on the Behavior of Families in Italy: A Focus on Children and Adolescents. *Front. Public Health* **2021**, *9*, 608358. [CrossRef] [PubMed]
13. Larsen, B.; Luna, B. Adolescence as a neurobiological critical period for the development of higher-order cognition. *Neurosci. Biobehav. Rev.* **2018**, *94*, 179–195. [CrossRef] [PubMed]
14. Best, O.; Ban, S. Adolescence: Physical changes and neurological development. *Br. J. Nurs.* **2021**, *30*, 272–275. [CrossRef] [PubMed]
15. Guessoum, S.B.; Lachal, J.; Radjack, R.; Carretier, E.; Minassian, S.; Benoit, L.; Moro, M.R. Adolescent psychiatric disorders during the COVID-19 pandemic and lockdown. *Psychiatr. Res.* **2020**, *291*, 113264. [CrossRef] [PubMed]
16. Ougrin, D.; Wong, B.H.; Vaezinejad, M.; Plener, P.L.; Mehdi, T.; Romaniuk, L.; Barrett, E.; Hussain, H.; Lloyd, A.; Tolmac, J.; et al. Pandemic-related emergency psychiatric presentations for self-harm of children and adolescents in 10 countries (PREP-kids): A retrospective international cohort study. *Eur. Child. Adolesc. Psychiatr.* **2022**, *31*, 1–13. [CrossRef] [PubMed]
17. Panchal, U.; Salazar de Pablo, G.; Franco, M.; Moreno, C.; Parellada, M.; Arango, C.; Fusar-Poli, P. The impact of COVID-19 lockdown on child and adolescent mental health: Systematic review. *Eur. Child. Adolesc. Psychiatry* **2023**, 32, 1151–1177. [CrossRef] [PubMed]
18. Theberath, M.; Bauer, D.; Chen, W.; Salinas, M.; Mohabbat, A.B.; Yang, J.; Chon, T.Y.; Bauer, B.A.; Wahner-Roedler, D.L. Effects of COVID-19 pandemic on mental health of children and adolescents: A systematic review of survey studies. *SAGE Open Med.* **2022**, *30*, 20503121221086712. [CrossRef]
19. Serra, G.; Apicella, M.; Iannoni, M.E.; Trasolini, M.; Andracchio, E.; Chieppa, F.; Averna, R.; Guidetti, C.; Maglio, G.; Reale, A.; et al. Urgent Psychiatric Consultations for Suicidal Ideation and Behaviors in Italian Adolescents during Different COVID-19 Pandemic Phases. *J. Pers. Med.* **2023**, *13*, 1112. [CrossRef]
20. National Collaborating Centre for Mental Health (UK). *Self-Harm: The Short-Term Physical and Psychological Management and Secondary Prevention of Self-Harm in Primary and Secondary Care*; National Institute for Health and Care Excellence: Guidelines; British Psychological Society (UK): Leicester, UK, 2004. Available online: http://www.ncbi.nlm.nih.gov/books/NBK56385/ (accessed on 1 January 2004).
21. American Psychiatric Association. *Diagnostic and Statistical Manual of Mental Disorders: DSM-5*, 5th ed.; American Psychiatric Association: Washington, DC, USA, 2013; 947p.
22. Kaufman, J.; Birmaher, B.; Rao, U.; Ryan, N. *K-SADS-PL DSM-5®. Intervista Diagnostica Per la Valutazione dei Disturbi Psicopatologici in Bambini e Adolescenti*; Erikson: Rome, Italy, 2019.
23. Sharpe, D. Chi-Square Test is Statistically Significant: Now What? *Pract. Assess Res. Evaluation* **2023**, *20*, 8.
24. Crawley, E.; Loades, M.; Feder, G.; Logan, S.; Redwood, S.; Macleod, J. Wider collateral damage to children in the UK because of the social distancing measures designed to reduce the impact of COVID-19 in adults. *BMJPO* **2020**, *4*, e000701. [CrossRef] [PubMed]

25. Munro, A.P.S.; Faust, S.N. Children are not COVID-19 super spreaders: Time to go back to school. *Arch Dis. Child.* **2020**, *105*, 618–619. [CrossRef] [PubMed]
26. Lazuras, L.; Barkoukis, V.; Tsorbatzoudis, H. Face-to-face bullying and cyberbullying in adolescents: Trans-contextual effects and role overlap. *Technol. Soc.* **2017**, *48*, 97–101. [CrossRef]
27. Viner, R.M.; Russell, S.J.; Croker, H.; Packer, J.; Ward, J.; Stansfield, C.; Mytton, O.; Bonell, C.; Booy, R. School closure and management practices during coronavirus outbreaks including COVID-19: A rapid systematic review. *Lancet Child. Adolesc. Health* **2020**, *4*, 397–404. [CrossRef]
28. Montreuil, M.; Camden, C.; Genest, C.; Gilbert, E.; Laberge-Perrault, E.; Piché, G.; Rassy, J.; Bogossian, A.; Gendron-Cloutier, L.; Barbo, G. Children and adolescents' mental health in pandemics and confinement: A scoping review of vulnerability factors and repercussions. *J. Child Health Care* **2023**, 136749352311655. [CrossRef] [PubMed]
29. Oliva, S.; Russo, G.; Gili, R.; Russo, L.; Di Mauro, A.; Spagnoli, A.; Alunni Fegatelli, D.; Romani, M.; Costa, A.; Veraldi, S.; et al. Risks and Protective Factors Associated With Mental Health Symptoms During COVID-19 Home Confinement in Italian Children and Adolescents: The #Understandingkids Study. *Front. Pediatr.* **2021**, *9*, 664702. [PubMed]
30. Amran, M.S. Psychosocial risk factors associated with mental health of adolescents amidst the COVID-19 pandemic outbreak. *Int. J. Soc. Psychiatry* **2022**, *68*, 6–8. [CrossRef]
31. Tang, W.C.; Lin, M.P.; You, J.; Wu, J.Y.W.; Chen, K.C. Prevalence and psychosocial risk factors of nonsuicidal self-injury among adolescents during the COVID-19 outbreak. *Curr. Psychol.* **2023**, *42*, 17270–17279. [CrossRef]
32. Mucci, M.; Lenzi, F.; D'Acunto, G.M.; Gazzillo, M.; Accorinti, I.; Boldrini, S.; Distefano, G.; Falcone, F.; Fossati, B.; Giurdanella Annina, R.; et al. How COVID-19 Phases Have Impacted Psychiatric Risk: A Retrospective Study in an Emergency Care Unit for Adolescents. *Children* **2022**, *9*, 1921. [CrossRef]
33. Plener, P.L. COVID-19 and Nonsuicidal Self-Injury: The Pandemic's Influence on an Adolescent Epidemic. *Am. J. Public Health* **2021**, *111*, 195–196. [CrossRef]
34. Christoffersen, M.N.; Møhl, B.; DePanfilis, D.; Vammen, K.S. Non-Suicidal Self-Injury—Does social support make a difference? An epidemiological investigation of a Danish national sample. *Child. Abuse Negl.* **2015**, *44*, 106–116. [CrossRef] [PubMed]
35. Xin, M.; Luo, S.; She, R.; Yu, Y.; Li, L.; Wang, S.; Ma, L.; Tao, F.; Zhang, J.; Zhao, J.; et al. Negative cognitive and psychological correlates of mandatory quarantine during the initial COVID-19 outbreak in China. *Am. Psychol.* **2020**, *75*, 607–617. [CrossRef] [PubMed]
36. Clarke, S.; Allerhand, L.A.; Berk, M.S. Recent advances in understanding and managing self-harm in adolescents. *F1000Research* **2019**, *8*, 1794. [CrossRef] [PubMed]
37. Olfson, M.; Wall, M.; Wang, S.; Crystal, S.; Bridge, J.A.; Liu, S.M.; Blanco, C. Suicide After Deliberate Self-Harm in Adolescents and Young Adults. *Pediatrics* **2018**, *141*, e20173517. [CrossRef]
38. Wilkinson, P.; Kelvin, R.; Roberts, C.; Dubicka, B.; Goodyer, I. Clinical and Psychosocial Predictors of Suicide Attempts and Nonsuicidal Self-Injury in the Adolescent Depression Antidepressants and Psychotherapy Trial (ADAPT). *Am. J. Psychiatry* **2011**, *168*, 495–501. [CrossRef]
39. Asarnow, J.R.; Porta, G.; Spirito, A.; Emslie, G.; Clarke, G.; Wagner, K.D.; Vitiello, B.; Keller, M.; Birmaher, B.; McCracken, J.; et al. Suicide Attempts and Nonsuicidal Self-Injury in the Treatment of Resistant Depression in Adolescents: Findings from the TORDIA Study. *J. Am. Acad. Child. Adolesc. Psychiatry* **2011**, *50*, 772–781. [CrossRef]
40. Bronfenbrenner, U. (Ed.) The bioecological theory of human development. In *Making Humans Being Human*; Sage: Thousand Oaks, CA, USA, 2005; pp. 3–15.
41. Zhou, X. Managing psychological distress in children and adolescents following the COVID-19 epidemic: A cooperative approach. *Psychol. Trauma* **2020**, *12*, S76–S78. [CrossRef]

children

MDPI

Article

Unpacking the Mood States of Children and Youth in Saskatchewan, Canada, in the Context of the COVID-19 Pandemic: Insights from the "See Us, Hear Us 2.0" Study

Nahin Shakurun [1,2], Tamara Hinz [3,4], Daniel A. Adeyinka [4] and Nazeem Muhajarine [1,2,*]

1 Department of Community Health and Epidemiology, College of Medicine, University of Saskatchewan, 107 Wiggins Road, Saskatoon, SK S7N 5E5, Canada; nahin.shakurun@usask.ca
2 Saskatchewan Population Health and Evaluation Research Unit (SPHERU), University of Saskatchewan, 104 Clinic Place, Saskatoon, SK S7N 2Z4, Canada
3 Department of Psychiatry, College of Medicine, University of Saskatchewan, 103 Hospital Drive, Saskatoon, SK S7N 0W8, Canada; tamara.hinz@usask.ca
4 Saskatchewan Health Authority, Royal University Hospital, Saskatoon, SK S7N 0W8, Canada; daniel.adeyinka@saskhealthauthority.ca
* Correspondence: nazeem.muhajarine@usask.ca; Tel.: +1-306-966-7940

Abstract: Background/Objectives: The COVID-19 pandemic created a growing need for insights into the mental health of children and youth and their use of coping mechanisms during this period. We assessed mood symptoms and related factors among children and youth in Saskatchewan. We examined if coping abilities mediated the relationship between risk factors and mood states. Methods: "See Us, Hear Us 2.0", a cross-sectional study of 563 child–parent dyads, provided the data. The dependent variable, current mood state, was measured by the CoRonavIruS health Impact Survey (CRISIS) scale. Independent variables included sociodemographics, behaviors, household conditions, and coping ability. Multiple linear regression and mediation analyses were conducted, ensuring sample representativeness with sampling weights and addressing missing data through multiple imputations. Results: The participants reported mood symptoms ("moderate" to "extreme") ranging from 23% to 38% on the CRISIS scale. Factors such as older children, hybrid learning, disrupted activities, and increased screen time worsened moods. The ethnic minority groups (BIPOC) living in mid-sized cities/towns experienced more negative moods compared to Whites residing in cities. Coping ability mediated the relationship between extracurricular activities and mood states. Conclusions: Our results underscore the importance of tailored interventions, recognizing the diverse needs of specific age groups, gender identities, and ethnicities and addressing the adverse effects of the pandemic-related disruptions on the mental health and well-being of school children in Saskatchewan. Our study also suggests prioritizing the diverse needs of children and youth during the planning and implementation of mental health services in the province.

Keywords: COVID-19; pandemic; post-COVID era; mood states; CRISIS scale; children and youth

Academic Editor: Matteo Alessio Chiappedi

Received: 13 November 2024
Revised: 7 January 2025
Accepted: 9 January 2025
Published: 10 January 2025

Citation: Shakurun, N.; Hinz, T.; Adeyinka, D.A.; Muhajarine, N. Unpacking the Mood States of Children and Youth in Saskatchewan, Canada, in the Context of the COVID-19 Pandemic: Insights from the "See Us, Hear Us 2.0" Study. *Children* **2025**, *12*, 79. https://doi.org/10.3390/children12010079

1. Introduction

In the post-COVID era, the mental health of children remains a crucial concern, as many are still struggling to cope with the psychological consequences of the pandemic. Stress and emotional instability have been intensified by disruptions in routine, prolonged isolation, and educational setbacks [1–4]. Although routines are returning to normal, some children still struggle to readjust to social settings and academic demands [5]. These challenges have left enduring effects on emotional and cognitive development, highlighting the importance of addressing mental health needs in this new landscape.

Several studies conducted during the pandemic revealed heightened mental health symptoms among children and youth [6,7]. A French study demonstrated that depression (29%), anxiety (49%), and irritability (51%) were more reported by 6- to 17-year-old children [8]. Hawke et al. [9] employed the CoRonavIruS health Impact Survey (CRISIS) scale (range: from 1 to 5) in their study, which specifically measures the symptoms of mental health challenges associated with COVID-19, and found a mean mood score of 3.14 (standard deviation = 0.77) for the 14–28-year-old community sample in Ontario, indicating a deterioration from the prepandemic level. A study of 8–18-year-old children and youth in Saskatchewan revealed that the point prevalence of medium to high anxiety and depression was 10.19% and 9.26%, respectively, in 2021 [10].

Early studies on COVID-19 found that mental health problems were more pronounced among older children and girls [11,12]. Mounting evidence suggests the link between urban–rural location and economic downturn to psychological well-being [10,13]. Similarly, a sudden shift to online or remote learning modalities exacerbated many psychological difficulties, e.g., anxiety, depression, inattention, loneliness, etc. [14–17]. Many studies have reported the impact of health behaviors (i.e., physical activity, sleep patterns, and screen use) on mental health symptoms [18–21]. A study of Iranian high school students revealed that moderate-to-vigorous physical activity was significantly associated with the negative moods such as anxiety, depression, and stress [22]. Another study revealed that more physical activity and less screen time were associated with better mental health for 6- to 17-year-old American children [23]. Additionally, economic downturn and interpersonal conflicts among family members potentially risk children to mental health adversities [24,25]. Longitudinal studies also reported that children and youth continue to suffer from psychological symptoms long after the COVID-19 pandemic [26,27]. The National Survey of Children's Health (NSCH) reported that about one in five adolescents (12–17 years old) in the United States had a diagnosed mental or behavioral condition in 2023 along with disengagement from school and difficulty in making social connections [28].

According to the Canada 2021 census profile, nearly 28% of the population in Saskatchewan was under 19 years of age [29]. Saskatchewan's unique population context warrants targeted analysis of children's mental health in the post-COVID-19 era due to its distinct demographic and geographical characteristics. With 35% of the population living in rural areas, vast geographical distances with many communities located hours away from major urban centers, and 16% of the population being Indigenous, Saskatchewan faces singular challenges in mental health service delivery [30]. The resource-based economy, higher rates of child poverty in certain regions, and limited access to specialized mental health services outside urban centers create additional vulnerabilities [31,32]. These factors, combined with COVID-19's disruption of social support systems and educational services and the government's particular response, necessitate a focused investigation of different dimensions of children's mental health along with potential modifiable factors that can provide the foundation for policies and interventions to strengthen provincial prevention measures.

The goal of this study was to report specific vulnerability factors of children and youth in Saskatchewan in the latter stages of the COVID-19 pandemic. To achieve this, our primary objective was to investigate different levels of mood symptoms present among children and youth at the end of the 2021–2022 academic year in Saskatchewan. The second objective was to provide evidence of associated factors with the mood states. We also analyzed the role of coping mechanisms as a potential mediator in the relationship between risk factors and negative mood states.

2. Materials and Methods

2.1. Study Design

We used data from "See Us, Hear Us (SUHU) 2.0", a cross-sectional study that was conducted between 19 May and 21 July 2022. Data were collected through an online survey questionnaire. At the time of data collection, schools were reopened, and all public health restrictions (e.g., mask use, physical distancing, or other restrictions from the 2020–2021 school year) were lifted for the 2021–2022 school year [33]. Also, vaccination was continued for children aged 12 years and above.

2.2. Participants

This study employed a hybrid sampling design: an online community panel and a voluntary, self-selected sample from the general population. The Canadian Hub for Applied and Social Research-developed Saskatchewan online Community Panel recruits its members through rigorous and robust probability-based sampling methods, which ensures a more representative sample that closely aligns with the general demographic make-up of the province of Saskatchewan. The second source, a self-selected sample from the eligible population, was directed to the CHASR's online survey. The invitation to participate in this study was sent by CHASR. To ensure the representativeness of the sample, weighting was applied using the iterative proportional method (also known as raking) based on age, gender, and location of residence using 2016 Canadian Census data for the targeted population [34].

A total of 563 child–parent dyads completed the online survey. In this study, the adult parent/caregiver of the child received the invitation to participate, not the child/youth directly. When there was more than one child/youth in the eligible age group present in the household (from 8 to 18 years), the child/youth whose birthday was closest to the date of survey was invited. This ensured one parent–child dyad per family or household participating in this study. These measures in place when recruiting would have had a countervailing effect to 'rebalance' any self-selection biases. Informed online consent was obtained before participation in the survey from both children and parents/caregivers. This study was approved by the Research Ethics Board of the University of Saskatchewan (Beh-2561).

2.3. Measures

2.3.1. Dependent Variable

The outcome variable was current mood states, measured by the CoRonavIruS health Impact Survey Child Mood States Scale. The CRISIS scale measures the following 8 items: worry, self-reported depression, self-reported anxiety, attentiveness, fatigue, fidgetiness, loneliness, and irritability [35]. Amongst the other available mental health assessment tools, the CRISIS scale was selected as a newly validated and widely adopted measure of child mental health, specifically for the COVID-19 pandemic. Children and youth were asked to rate their moods on a 5-point Likert-type scale during the past two weeks preceding data collection (therefore referred to as "current mood states"). Items were recorded and averaged to generate a total score ranging from 0 to 4, with the high score indicating worse/negative moods. The categories of the items were collapsed to describe the prevalence of the components and reflect a mood symptom present at least moderately, compared to being "not at all" or "slightly". Later, principal component analysis (PCA) was performed with the original items of the scale. The suitability of the items for data reduction was tested by Bartlett's test of sphericity (p-value < 0.001), and the Kaiser–Meyer–Olkin

(KMO) test for sampling adequacy exhibited a high adequacy level above 0.8 [36]. The principal component was retained with an eigenvalue >1, and that explained 100% of the variability of the CRISIS scale. The score for the principal component was generated and treated as a continuous outcome variable for further analysis. Throughout this paper, the CRISIS scale and the PCA score were interpreted in a manner where a lower score indicates a better mood and a higher score indicates a worse negative mood. The internal consistency of the measure was good for the current mental health of Canadian youth (Cronbach's alpha = 0.88) [9].

2.3.2. Independent Variables

Guided by the concept of Health Canada's "Population Health Framework" (PHF) [37] and published articles, a set of independent variables was selected for this analysis. We considered learning modalities (in-person schooling, a mix of online and in-person, and online/homeschooling only) and any change in extracurricular activities as potential lifestyle factors. Any change in behavioral factors like physical activity, screen time, and sleep pattern in the past month before the data collection was assessed. Household financial stability and conflicts in the family were evaluated. We also considered the presence of positive COVID-19 cases at home, infection among acquaintances, and the coping capacity of the children and youth. Demographic information was collected for age, grade, gender, ethnicity, gross household income, location of residence, and parental immigration status. Details of these variables with their categorization and conceptual framework for this study are presented in the Supplementary Table (Figure S1 and Table S1).

2.4. Statistical Analysis

Data were analyzed using STATA statistical software version 17 [38]. Overall, the missing values ranged from 0% to 10%. We treated the "do not know or prefer not to answer" as missing for the income variable, so the missing values increased to 17%. Multiple imputation by chained equation technique was applied to address the missing data, and the pooled results were reported throughout this manuscript [39].

The sample characteristics were presented with frequencies and percentages. Bivariate analysis was performed between each independent variable and the outcome using simple linear regression. A p-value of <0.25 was used to select the candidate variables for multiple linear regression analysis. Multicollinearity among the selected variables was assessed by the mean variance inflation factor (VIF). Finally, the findings from the multivariable linear regression analysis with the forward selection approach were reported with unstandardized regression coefficients (b) and corresponding 95% confidence intervals (CI). A p-value of less than 0.05 was considered significant. Pre-selected interactions were assessed to determine the association of current mood states with age, gender, and ethnicity that changed across the location of residence, income levels, and immigration categories. Lastly, to address our third aim, the mediator role of children's coping capacity was assessed between risk factors and the mood states [40]. The risk factors included in this study were selected based on published articles and on the advice from the child–parent advisory council and our clinical (child psychiatry) experience and insights.

3. Results

3.1. Sample Characteristics

Table 1 represents the demographic characteristics of our respondents. About 41.1% of the participants were 8–11 years old, 34.7% were 12–14 years old, and 24.1% were 15–18 years old. Over half of the respondents were in elementary grades (Grade 1–8) (63.4%). Approximately 48.0% of the children and youth self-identified as a boy, 47.4% as a girl, and 4.5% as other (neither boy nor girl). Regarding ethnocultural identity, the majority reported being White (79.6%), while one-fifth of the respondents identified as BIPOC (Black, Indigenous, and People of Color) (20.3%). Almost half of the respondents were from cities (Saskatoon/Regina) (49.9%). About 58.3% of respondents reported a yearly household income (pre-tax) of CAD 100,000 or higher, and about 15.5% of the respondents reported migrant status. Based on the census data of 2021, our sample is well representative of the targeted population of Saskatchewan [29].

Table 1. Demographic characteristics of the participants (n = 563); "See Us, Hear Us 2.0" study, Saskatchewan, Canada.

Characteristics	Frequency (%)	Imputed and Weighted Frequency (%)
Age		
8–11 years	246 (43.69)	232 (41.19)
12–15 years	198 (35.17)	195 (34.70)
16–18 years	119 (21.14)	136 (24.11)
Missing	0	
Grade		
Elementary	355 (63.06)	357 (63.47)
High	180 (31.97)	206 (36.53)
Missing	28 (4.97)	
Gender		
Boy	286 (50.80)	271 (48.06)
Girl	247 (43.87)	267 (47.43)
Others	25 (4.44)	25 (4.51)
Missing	5 (0.89)	
Ethnicity		
White	441 (78.33)	449 (79.69)
BIPOC	117 (20.78)	114 (20.31)
Missing	5 (0.89)	
Gross household income		
Less than CAD 100,000	197 (34.99)	235 (41.67)
CAD 100,000 or more	268 (47.60)	328 (58.33)
Missing	98 (17.41)	
Location of residence		
Saskatoon/Regina	265 (47.07)	281 (49.95)
Mid-sized city/town	49 (8.70)	88 (15.55)
Rural	192 (34.10)	194 (34.50)
Missing	57 (10.12)	
Parent immigration status		
Both born in Canada	461 (81.88)	476 (84.48)
Neither/at least one parent born in Canada	94 (16.70)	87 (15.52)
Missing	8 (1.42)	

3.2. Mood Symptoms and Associated Factors

In the 2021–2022 school year, children and youth reported varying levels of mood symptoms. Figure 1 illustrates the prevalence of eight mood symptoms as measured by the CRISIS scale. The prevalence of moderate to severe mood changes ranged from 23% to38%. The prevalence of moderate to extreme irritability (38.4%) and fatigue (38.0%) was higher, whereas the prevalence of moderate to severe depression was the lowest (23.3%).

A detailed distribution of the different categories of the CRISIS items is presented in the Supplementary File (Table S2).

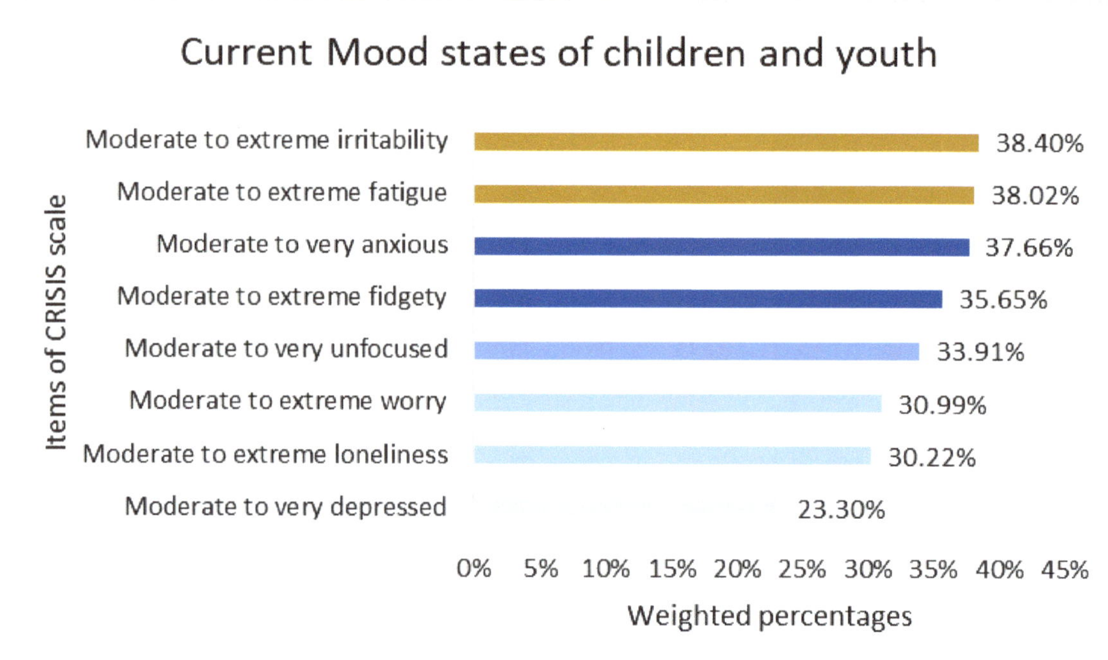

Figure 1. Prevalence of mood symptoms (from "moderate" to "extreme" in the CRISIS scale) in children and youth (8–18 years) in Saskatchewan.

Table 2 shows the coefficients from univariate (unadjusted) and multiple regression models (adjusted). In the adjusted model, negative moods were significantly higher among 16–18-year-old children compared to 8–11-year-olds (b = 0.27; 95% CI: 0.02, 0.52; $p < 0.05$). Children who did not self-identify as boys/girls had 0.78 points higher mood scores than those who self-identified as boys (95% CI: 0.35, 1.21; $p < 0.001$). According to the location of residence, children residing in rural areas had significantly lower mood scores compared to those from the cities (Saskatoon/Regina) (b = −0.17; 95% CI: −0.34, 0.00; $p < 0.05$).

Table 2. Unadjusted and adjusted regression models for factors associated with the mood states score of children and youth in Saskatchewan, Canada (results are presented as unstandardized coefficients with 95% confidence intervals).

Characteristics	Unadjusted Coefficient (95% CI)	Adjusted Coefficient (95% CI)
Age		
8–11 years	Ref	Ref
12–15 years	0.23 (0.04, 0.42) *	0.11 (−0.05, 0.26)
16–18 years	0.43 (0.21, 0.66) **	0.27 (0.02, 0.52) *
Grade		
Elementary	Ref	Ref
High	0.23 (0.04, 0.41) *	−0.01 (−0.21, 0.19)

Table 2. *Cont.*

Characteristics	Unadjusted Coefficient (95% CI)	Adjusted Coefficient (95% CI)
Gender		
Boy	Ref	Ref
Girl	0.13 (−0.04, 0.31)	0.10 (−0.03, 0.23)
Others	0.93 (0.45, 1.40) **	0.78 (0.35, 1.21) **
Ethnicity		
White	Ref	Ref
BIPOC	0.06 (−0.17, 0.29)	0.05 (−0.27, 0.37)
Gross household income		
Less than CAD 100,000	Ref	Ref
CAD 100,000 or more	−0.25 (−0.43, −0.07) *	−0.07 (−0.24, 0.10)
Location of residence		
Saskatoon/Regina	Ref	Ref
Mid-sized city/town	0.36 (0.06, 0.66) *	0.07 (−0.18, 0.32)
Rural	−0.18 (−0.36, −0.01) *	−0.17 (−0.34, −0.00) *
Parent immigration status		
Both born in Canada	Ref	Ref
Neither/at least one parent born in Canada	−0.11 (−0.33, 0.11)	0.24 (−0.07, 0.55)
Learning method		
Attended in-person all the year	Ref	Ref
Mix of online and in-class learning	0.43 (0.18, 0.68) *	0.24 (0.04, 0.43) *
Online/Others	0.31 (−0.15, 0.76)	0.09 (−0.23, 0.41)
Impact of pandemic on extracurricular activities		
No impact	Ref	Ref
A little or a lot of impact	0.39 (0.22, 0.57) **	0.18 (0.04, 0.33) *
Change in physical activity		
More active	Ref	Ref
The same	0.28 (0.06, 0.50) *	0.21 (0.05, 0.37) *
Less active	0.47 (0.22, 0.71) **	0.15 (−0.08, 0.38)
Change in sleep pattern		
Better	Ref	Ref
The same	−0.23 (−0.41, −0.05) *	−0.10 (−0.25, 0.05)
Worse	0.52 (0.17, 0.87) *	0.21 (−0.10, 0.52)
Change in screen time		
Decreased	Ref	Ref
The same	0.16 (−0.02, 0.34)	0.07 (−0.09, 0.23)
Increased	0.66 (0.43, 0.89) **	0.34 (0.16, 0.53) **
COVID-19/COVID-like cases at home		
No cases	Ref	Ref
Positive case was present	0.29 (0.11, 0.48) *	0.20 (0.04, 0.36) *
Severity outside household		
No	Ref	Ref
Yes, severely ill/died	0.22 (−0.02, 0.46)	0.10 (−0.09, 0.28)
Financial stability		
Secure	Ref	Ref
Insecure	0.29 (0.10, 0.48) *	0.05 (−0.10, 0.21)

Table 2. *Cont.*

Characteristics	Unadjusted Coefficient (95% CI)	Adjusted Coefficient (95% CI)
Coping ability		
Family conflict		
None/somewhat less	Ref	Ref
No real change	−0.05 (−0.38, 0.29)	−0.07 (−0.38, 0.24)
Varied	0.24 (−0.11, 0.58)	0.06 (−0.26, 0.38)
A lot/somewhat more	0.59 (0.26, 0.92) **	0.32 (0.00, 0.64) *
Most times/always	Ref	Ref
Sometimes	0.65 (0.50, 0.81) **	0.48 (0.34, 0.63) **
Hardly ever	0.97 (0.60, 1.34) **	0.70 (0.40, 0.99) **

N.B. * Significance at $p < 0.05$; ** Significance at $p < 0.001$.

Among the factors related to daily life, hybrid learning modalities (b = 0.24; 95% CI: 0.04, 0.43; $p < 0.05$) and disrupted extracurricular activities (b = 0.18; 95% CI: 0.04, 0.33; $p < 0.05$) were significantly associated with higher mood scores in children and youth during the second academic year into the pandemic. Those who had "the same" amount of physical activities as earlier in the pandemic had scored more on the CRISIS scale compared to those who had increased activity (b = 0.21; 95% CI: 0.05, 0.37; $p < 0.05$). Likewise, increased screen time in the past month was significantly associated with increased mood score, 0.34 coefficient points higher, compared to those whose screen time had decreased (b = 0.34; 95% CI: 0.16, 0.53; $p < 0.001$). Additionally, conflicts among the family members were significantly associated with increased mood scores (a lot/somewhat more, compared to none/somewhat less; b = 0.32; 95% CI: 0.00, 0.64; $p < 0.05$). Children who said that they had less capacity to cope experienced significantly higher mood scores during this period (sometimes, b = 0.48; 95% CI: 0.34, 0.63; $p < 0.001$ and hardly ever, b = 0.70; 95% CI: 0.40, 0.99; $p < 0.001$).

We found a significant moderating effect involving ethnicity and location of residence. (Figure 2a). BIPOC children and youth were associated with lower mood scores, compared to white counterparts, for those who resided in urban centers or in rural communities. This association was significantly modified—in fact, reversed, or heightened—for those who lived in mid-sized communities. Another significant moderating effect involved ethnicity and immigration status. BIPOC children whose parents were Canadian-born were associated with a probability of higher mood scores compared to BIPOC children with at least one parent born outside of Canada (Figure 2b).

Figure 3 represents the mediational effect of coping ability. The effect of disrupted extracurricular activities on mood states of children and youth is significantly mediated through coping ability (indirect effect, $a \times b = 2.76$; $p < 0.05$). The ratio of indirect effect to total effect (RIT = $(a \times b)/C$) was 0.94, i.e., 94% of the total effect of disrupted extracurricular activities on the current mood states of children and youth in Saskatchewan was mediated by the respondent's coping ability.

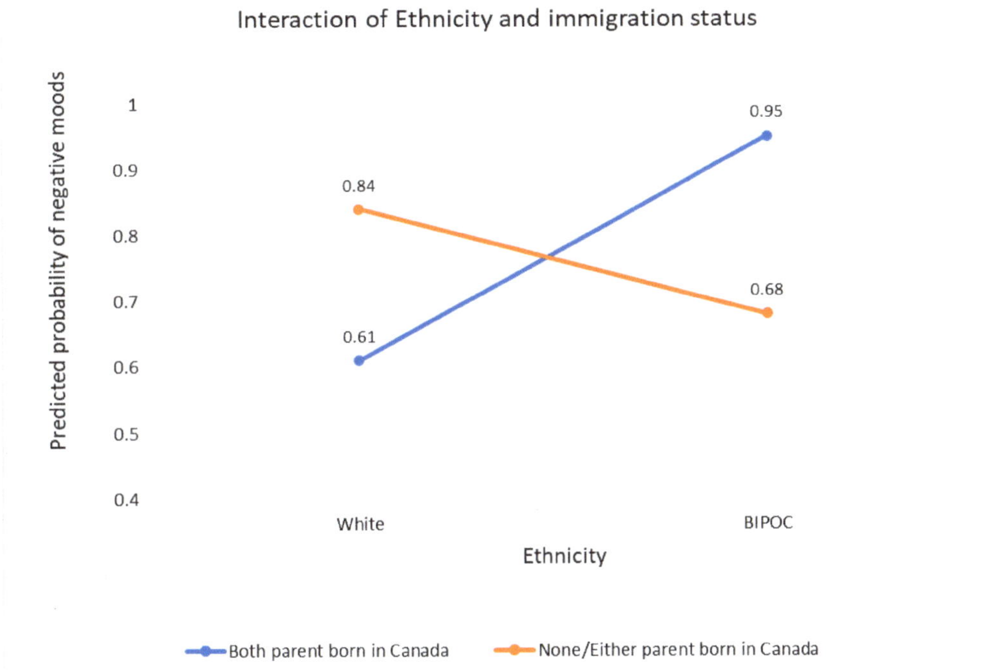

(a)

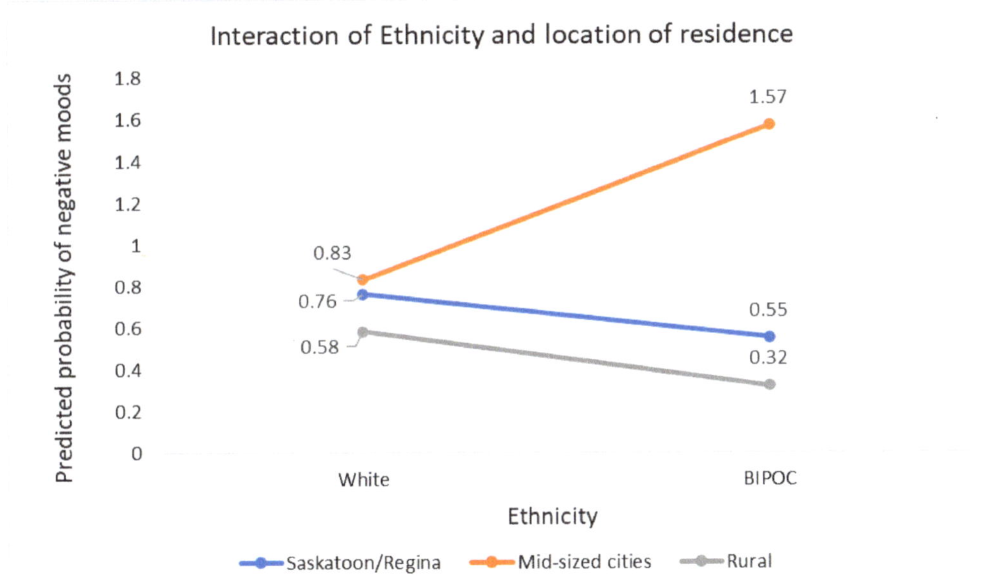

(b)

Figure 2. (**a**) Location of residence modifies the effect of ethnicity in predicting mood states in children and youth in Saskatchewan. (**b**) Immigration status modifies the effect of ethnicity in predicting negative mood states in children and youth in Saskatchewan.

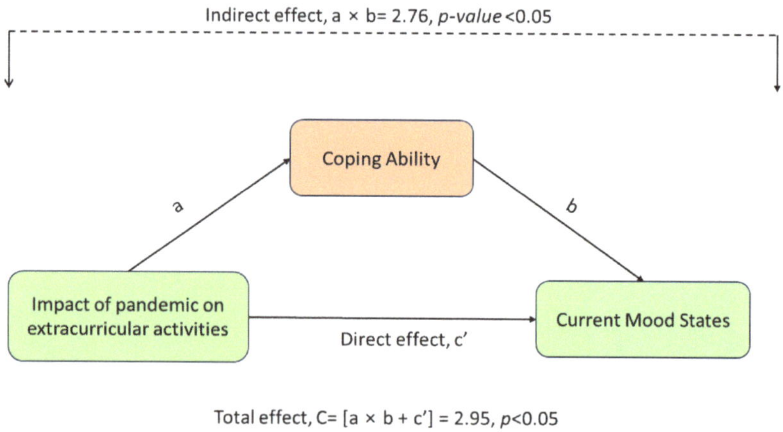

Figure 3. The mediational effect of coping ability of children and youth in the relationship between extracurricular activities and the current mood states of the respondents.

4. Discussion

The present study assessed the different levels of mood states and their correlates in children and youth using data from the "See Us, Hear Us 2.0" study in Saskatchewan, Canada. This study identified a significant mediational path to mood states through the coping ability of children and youth. We found that children and youth in Saskatchewan experienced varying levels of mood symptoms according to the CRISIS scale during the 2021–2022 academic year. More than one-third of the respondents had moderate to extreme irritability (38.40%) and fatigue (38.02%) among the eight components of mood symptoms, which are comparable with the previously published articles [41–43].

We found respondents aged 16 to 18 years reported higher mood scores, which is aligned with the findings from a study indicating increased emotional issues such as sadness and irritability among 9470 adolescents aged between 12 and 17 years old [44]. Conversely, during the first academic year (2020–2021), Saskatchewan children aged 8–11 experienced heightened anxiety and depression [10]. This shift from younger to older children may stem from biological, developmental (e.g., hormonal changes), and psychosocial factors (e.g., academic and interpersonal stresses) [45]. When comparing the mood scores across genders, there was an association between higher mood scores and respondents who did not identify themselves as boy or girl (such as those identifying as non-binary, two-spirit, or others). However, previous studies report results in the opposite direction. [25,44]. An Australian research emphasized increased depressive and anxiety symptoms in girls [46]. However, this disparity compels us to prioritize equity-deserving groups like sexual and gender minorities and underscore the necessity for tailored mental health support for them.

Similar to other studies, our research indicates that residential location has an impact on mental health [47,48]. We found an association between rural children and better mood states, compared to those in cities (Saskatoon/Regina) and mid-sized towns. It is possible that this reflects to a degree the relatively greater disruption of normal life and routine in cities compared to rural communities during the pandemic. Greater access to nature, lower-density population centers and events and reduced environmental stressors may have also contributed to the relative diminishment of poor mental health symptoms in rural children and youth. Additionally, strong community ties and support in rural areas may have aided in overcoming mental health challenges. However, further exploration with detailed qualitative components is required to understand the extent to which any social stigma or cultural norms may have hindered children and youth in rural areas reporting or accessing mental health services.

In addition to sociodemographic factors, this study also highlighted the diverse effects of the pandemic on lifestyle and behavioral factors. Our study participants who had hybrid learning settings reported higher mood scores compared to those who attended school in person. Prior research has reported the detrimental effects of online learning on mental health early in the pandemic [15,46]. Our findings emphasize that even after public health restrictions were lifted, children in hybrid learning during the second academic year in the pandemic struggled to return to normalcy. The current study examined the link between extracurricular activities and mood states among children and youth in Saskatchewan and found that respondents with disrupted extracurricular activities since the onset of the pandemic had higher mood scores. This aligns with the findings from prior studies carried out among adolescents in various Canadian provinces [49,50]. Oberle et al. [49] revealed that extracurricular activities are linked to increased life satisfaction and decreased anxiety and depression. Likewise, LaForge-MacKenzie et al. [50] found that extracurricular activities improved mental health outcomes for Ontario's children and youth before and during the pandemic. We also identified that elevated screen time was responsible for increased mood scores, as observed in earlier pandemic studies [44,51,52]. Limited outdoor options during the pandemic's onset exacerbated screen dependency.

About 33.18% of respondents experienced increased family conflicts during the second year of the pandemic, significantly linked to increased mood symptoms, consistent with SUHU 1.0 findings [10] and Australian research [46]. The latter observed more depressive symptoms in children with heightened conflicts with their fathers, whereas conflicts with siblings, friends, or mothers did not impact anxiety or depressive symptoms [46]. Family dynamics, particularly parent–child relationships, crucially influence children's psychological well-being [53,54], especially during the pandemic when there was limited peer support available. Dysfunctional parenting processes, parental work–life conflicts, caregivers' mental health, and financial strain exacerbated familial chaos or conflicts during COVID-19 [54–57].

In this research, we found that children and youth who had less ability to cope expressed higher mood scores. Vallejo-Slocker et al. [58] demonstrated that nonactive coping predicted worse mental health, whereas problem-solving served as a protective factor for Spanish children. Zhang et al. [57] showed positive coping style, such as positive appraisal and thinking, distancing, problem-solving, and help-seeking, was protective against depression, anxiety, and trauma-related distress. These findings were generally in line with our results. The assessment of the various coping strategies adopted by the respondents was beyond the scope of this study. Further detailed evaluation of coping methods adopted by the children and youth is required, preferably through qualitative investigations.

This study revealed significant interactions between ethnocultural identity, location of residence, and parental immigrant status in Saskatchewan. BIPOC children in mid-sized cities exhibited notably higher mood scores compared to their counterparts in cities (Saskatoon/Regina). Moreover, there is an association between BIPOC children and youth whose parents were Canadian-born and significantly higher mood scores compared to those whose parents were born outside of Canada. Our findings provided insights into the unique challenges faced by this group, highlighting persistent disparities among socially disadvantaged groups. During the first year of the pandemic, BIPOC children from low-income families were more likely to report low-to-moderate quality of life compared to higher-income counterparts [10]. These findings emphasize the necessity for specialized mental health services tailored to marginalized populations in Saskatchewan.

The mediational effect of coping ability was a unique finding of this study. Our findings identified a significant mediational path between extracurricular activities and mood states among children and youth with significant indirect and total effects through their coping ability. The potential connection between disrupted extracurricular activities and negative mood experiences as a coping mechanism in children and youth remains underexplored in the literature. Few studies have specifically examined the relationship between participation in extracurricular activities and the development of practical coping abilities in children [59,60]. Creative activities, such as music and art, keeping social connections, routing building and engaging in physical activities were proven as effective coping strategies for emotional wellbeing and fostering resilience during the pandemic [31]. A longitudinal study of 1162 Australian children demonstrated that participation in extracurricular activities predicted improved coping efficacy [61]. Effective coping skills have been associated with various positive outcomes for adolescents, including higher self-esteem, reduced symptoms of depression and conduct problems, and enhanced social and academic competence [59]. Further qualitative studies are essential for a more comprehensive understanding of the coping strategies adopted by the children and youth in the post-pandemic period.

We incorporated the eight components of mood by the principal component analysis, which allowed us to reduce the dimensionality of the data but retain maximum variability of the components. Missing data were addressed using appropriate techniques, which made our findings more informative and valid. Another strength of this study was to elucidate the mediating effect of coping ability on the relationship between extracurricular activities and mood state among Saskatchewan's children and youth. As in any study, however, there are several limitations that should be noted. The cross-sectional survey data precluded revealing causal relationships. Although future longitudinal studies may tease out causality, cross-sectional studies have a place. They could reveal understanding of a problem, such as mental health, in real time and especially during an unprecedented pandemic. Due to multiple imputation methods, we were not able to test the separate path coefficients and establish the standardized mediational effect. Additionally, public health restrictions varied from province to province. Therefore, our results may not be generalizable in other jurisdictions.

5. Conclusions

Our study provides significant evidence on the magnitude of mood states of the children and youth, along with the associated factors, at the end of the second academic year since the pandemic began. The time when the data were collected, all the public health mandates were lifted, and schools resumed in-person classes. However, our study confirms the multidimensional mood experiences of children and youth in Saskatchewan at the end of the second academic year since the pandemic began. Our study also emphasized the significant sociodemographic and lifestyle factors that can inform policy changes and the reallocation of mental health resources to the equity-seeking groups while eliminating mental health disparities. Also, future research with qualitative components is warranted to understand the coping mechanism employed by the children and youth during this devastating situation. The collaboration between the researchers, mental health care providers, and policymakers is essential to design and implement targeted interventions for children and youth.

Supplementary Materials: The following supporting information can be downloaded at: https://www.mdpi.com/article/10.3390/children12010079/s1. Figure S1: Conceptual framework adapted from Population Health Framework (Diagram courtesy: Dr. William Pickett); Table S1: Independent variables and their categorization; Table S2: Distribution of mood states with missing values (N = 563).

Author Contributions: Conceptualization: N.S. and N.M.; methodology: N.S. and N.M.; formal analysis: N.S., D.A.A. and N.M.; writing—original draft: N.S.; writing—review and editing: N.S., T.H., D.A.A. and N.M.; supervision: N.M. All authors have read and agreed to the published version of the manuscript.

Funding: From the Canadian Institutes of Health Research (CIHR), Operating Grant: Understanding and mitigating impacts of COVID-19 on children, youth, and families, application #468606: See Us, Hear Us 2.0: Mental health and well-being impacts of the COVID-19 pandemic on children, youth, and families in Saskatchewan.

Institutional Review Board Statement: This study was approved by the University of Saskatchewan Research Ethics Board (Beh-2561).

Informed Consent Statement: Participants completed an online consent process after they had read information online outlining the purpose of this study, confidentiality, risks, benefits of participation, incentives, and data management.

Data Availability Statement: The data presented in this study are available upon reasonable request.

Acknowledgments: The authors would like to acknowledge the contribution of the parent and youth dyads, our community panel, at the development stage of the survey. We acknowledge the collaborators: Jane McLeod, Natalie Kallio, Isabelle Dena, Jessica McCutcheon, Nuvelle Novik, Tom McIntosh, and Akela Peoples. We thank the collaboration of the Saskatoon school divisions and the Saskatchewan League of Educational Administrators, Directors, and Superintendents (SaskLEADS) in this research. Our most sincere thanks to all children, youth, and their parents who participated in this study.

Conflicts of Interest: The authors declare no conflicts of interest.

References

1. Golberstein, E.; Wen, H.; Miller, B.F. Coronavirus Disease 2019 (COVID-19) and Mental Health for Children and Adolescents. *JAMA Pediatr.* **2020**, *174*, 819–820. [CrossRef] [PubMed]
2. Marques de Miranda, D.; da Silva Athanasio, B.; Sena Oliveira, A.C.; Simoes-e-Silva, A.C. How Is COVID-19 Pandemic Impacting Mental Health of Children and Adolescents? *Int. J. Disaster Risk Reduct.* **2020**, *51*, 101845. [CrossRef] [PubMed]
3. Mourouvaye, M.; Bottemanne, H.; Bonny, G.; Fourcade, L.; Angoulvant, F.; Cohen, J.F.; Ouss, L. Association between Suicide Behaviours in Children and Adolescents and the COVID-19 Lockdown in Paris, France: A Retrospective Observational Study. *Arch. Dis. Child.* **2021**, *106*, 918–919. [CrossRef] [PubMed]
4. Government of Canada, Statistics Canada. The Daily—Canadian Health Survey on Children and Youth, 2019. Available online: https://www150.statcan.gc.ca/n1/daily-quotidien/200723/dq200723a-eng.htm (accessed on 12 September 2023).
5. Minihan, S.; Orben, A.; Songco, A.; Fox, E.; Ladouceur, C.D.; Mewton, L.; Moulds, M.; Pfeifer, J.H.; Harmelen, A.-L.V.; Schweizer, S. Social Determinants of Mental Health during a Year of the COVID-19 Pandemic. *Dev. Psychopathol.* **2023**, *35*, 1701–1713. [CrossRef] [PubMed]
6. Panchal, U.; Salazar de Pablo, G.; Franco, M.; Moreno, C.; Parellada, M.; Arango, C.; Fusar-Poli, P. The Impact of COVID-19 Lockdown on Child and Adolescent Mental Health: Systematic Review. *Eur. Child Adolesc. Psychiatry* **2023**, *32*, 1151–1177. [CrossRef]
7. Sarkadi, A.; Sahlin Torp, L.; Pérez-Aronsson, A.; Warner, G. Children's Expressions of Worry During the COVID-19 Pandemic in Sweden. *J. Pediatr. Psychol.* **2021**, *46*, 939–949. [CrossRef]
8. Dönmez, Y.E.; Uçur, Ö. Frequency of Anxiety, Depression, and Irritability Symptoms in Children During the COVID-19 Outbreak and Potential Risk Factors Associated With These Symptoms. *J. Nerv. Ment. Dis.* **2021**, *209*, 727–733. [CrossRef]
9. Hawke, L.D.; Barbic, S.P.; Voineskos, A.; Szatmari, P.; Cleverley, K.; Hayes, E.; Relihan, J.; Daley, M.; Courtney, D.; Cheung, A.; et al. Impacts of COVID-19 on Youth Mental Health, Substance Use, and Well-Being: A Rapid Survey of Clinical and Community Samples: Répercussions de La COVID-19 Sur La Santé Mentale, l'utilisation de Substances et Le Bien-Être Des Adolescents: Un Sondage Rapide d'échantillons Cliniques et Communautaires. *Can. J. Psychiatry Rev. Can. Psychiatr.* **2020**, *65*, 701–709. [CrossRef]
10. Muhajarine, N.; Pisolkar, V.; Hinz, T.; Adeyinka, D.A.; McCutcheon, J.; Alaverdashvili, M.; Damodharan, S.; Dena, I.; Jurgens, C.; Taras, V.; et al. Mental Health and Health-Related Quality of Life of Children and Youth during the First Year of the COVID-19 Pandemic: Results from a Cross-Sectional Survey in Saskatchewan, Canada. *Children* **2023**, *10*, 1009. [CrossRef]

11. Esposito, S.; Giannitto, N.; Squarcia, A.; Neglia, C.; Argentiero, A.; Minichetti, P.; Cotugno, N.; Principi, N. Development of Psychological Problems Among Adolescents During School Closures Because of the COVID-19 Lockdown Phase in Italy: A Cross-Sectional Survey. *Front. Pediatr.* **2021**, *8*, 628072. [CrossRef]
12. Halldorsdottir, T.; Thorisdottir, I.E.; Meyers, C.C.A.; Asgeirsdottir, B.B.; Kristjansson, A.L.; Valdimarsdottir, H.B.; Allegrante, J.P.; Sigfusdottir, I.D. Adolescent Well-Being amid the COVID-19 Pandemic: Are Girls Struggling More than Boys? *JCPP Adv.* **2021**, *1*, e12027. [CrossRef] [PubMed]
13. McKune, S.L.; Acosta, D.; Diaz, N.; Brittain, K.; Beaulieu, D.J.-; Maurelli, A.T.; Nelson, E.J. Psychosocial Health of School-Aged Children during the Initial COVID-19 Safer-at-Home School Mandates in Florida: A Cross-Sectional Study. *BMC Public Health* **2021**, *21*, 603. [CrossRef] [PubMed]
14. Mridul1, B.B. Online Classes during COVID-19 Pandemic: Anxiety, Stress & Depression among University Students. *Indian J. Forensic Med. Toxicol.* **2021**, *15*, 186–189. [CrossRef]
15. Villones, J.M.; Dumlao, J.P., II; Maaño, M.M.C. The Incidence of Anxiety and Depression Among Adolescents (Age 15–18 Years Old) After 1 Year of Online Learning. *Asia Pac. J. Pediatr. Child Health* **2022**, *5*, 9–25.
16. Walters, T.; Simkiss, N.J.; Snowden, R.J.; Gray, N.S. Secondary School Students' Perception of the Online Teaching Experience during COVID-19: The Impact on Mental Wellbeing and Specific Learning Difficulties. *Br. J. Educ. Psychol.* **2022**, *92*, 843–860. [CrossRef]
17. Kılınçel, Ş.; Kılınçel, O.; Muratdağı, G.; Aydın, A.; Usta, M.B. Factors Affecting the Anxiety Levels of Adolescents in Home-quarantine during COVID-19 Pandemic in Turkey. *Asia-Pac. Psychiatry* **2021**, *13*, e12406. [CrossRef]
18. Zhao, J.; Xu, J.; He, Y.; Xiang, M. Children and Adolescents' Sleep Patterns and Their Associations with Mental Health during the COVID-19 Pandemic in Shanghai, China. *J. Affect. Disord.* **2022**, *301*, 337–344. [CrossRef]
19. Bothe, K.; Schabus, M.; Eigl, E.-S.; Kerbl, R.; Hoedlmoser, K. Self-Reported Changes in Sleep Patterns and Behavior in Children and Adolescents during COVID-19. *Sci. Rep.* **2022**, *12*, 20412. [CrossRef]
20. Marciano, L.; Viswanath, K.; Morese, R.; Camerini, A.-L. Screen Time and Adolescents' Mental Health before and after the COVID-19 Lockdown in Switzerland: A Natural Experiment. *Front. Psychiatry* **2022**, *13*, 981881. [CrossRef]
21. Alves, J.M.; Yunker, A.G.; DeFendis, A.; Xiang, A.H.; Page, K.A. Associations between Affect, Physical Activity, and Anxiety Among US Children During COVID-19. *medRxiv* **2020**, *23*, 1–25. [CrossRef]
22. Ghorbani, S.; Afshari, M.; Eckelt, M.; Dana, A.; Bund, A. Associations between Physical Activity and Mental Health in Iranian Adolescents during the COVID-19 Pandemic: An Accelerometer-Based Study. *Children* **2021**, *8*, 1022. [CrossRef] [PubMed]
23. Tandon, P.S.; Zhou, C.; Johnson, A.M.; Gonzalez, E.S.; Kroshus, E. Association of Children's Physical Activity and Screen Time With Mental Health During the COVID-19 Pandemic. *JAMA Netw. Open* **2021**, *4*, e2127892. [CrossRef] [PubMed]
24. Hanno, E.C.; Cuartas, J.; Miratrix, L.W.; Jones, S.M.; Lesaux, N.K. Changes in Children's Behavioral Health and Family Well-Being During the COVID-19 Pandemic. *J. Dev. Behav. Pediatr.* **2022**, *43*, 168. [CrossRef]
25. Waite, P.; Pearcey, S.; Shum, A.; Raw, J.A.L.; Patalay, P.; Creswell, C. How Did the Mental Health Symptoms of Children and Adolescents Change over Early Lockdown during the COVID-19 Pandemic in the UK? *JCPP Adv.* **2021**, *1*, e12009. [CrossRef]
26. Kaman, A.; Devine, J.; Wirtz, M.A.; Erhart, M.; Boecker, M.; Napp, A.-K.; Reiss, F.; Zoellner, F.; Ravens-Sieberer, U. Trajectories of Mental Health in Children and Adolescents during the COVID-19 Pandemic: Findings from the Longitudinal COPSY Study. *Child Adolesc. Psychiatry Ment. Health* **2024**, *18*, 89. [CrossRef]
27. Larsen, L.; Schauber, S.K.; Holt, T.; Helland, M.S. Longitudinal Covid-19 Effects on Child Mental Health: Vulnerability and Age Dependent Trajectories. *Child Adolesc. Psychiatry Ment. Health* **2023**, *17*, 104. [CrossRef]
28. Sppenfield, O.; Alberto, C.; Minnaert, J.; Donney, J.; Lebrun-Harris, L.; Ghandour, R. National Survey of Children's Health Adolescent Mental and Behavioral Health, 2023 Data Brief | October 2024. 2024. Available online: https://mchb.hrsa.gov/sites/default/files/mchb/data-research/nsch-data-brief-adolescent-mental-behavioral-health-2023.pdf (accessed on 4 November 2024).
29. Government of Canada, Statistics Canada. Profile Table, Census Profile, 2021 Census of Population—Saskatchewan [Province]. Available online: https://www12.statcan.gc.ca/census-recensement/2021/dp-pd/prof/index.cfm?Lang=E (accessed on 1 January 2025).
30. Government of Saskatchewan. Demography, Census Reports and Statistics | Saskatchewan Bureau of Statistics. Available online: https://www.saskatchewan.ca/government/government-data/bureau-of-statistics/population-and-census (accessed on 25 October 2023).
31. Pisolkar, V.; Dena, I.; Green, K.L.; Palmer-Clarke, Y.; Hinz, T.; Muhajarine, N. See Us, Hear Us! Children, Adolescents and Families in Saskatchewan Coping with Mental Health During the COVID-19 Pandemic. *Int. J. Qual. Stud. Health Well-Being* **2024**, *19*, 2361494. [CrossRef]
32. Sanchez, M. Saskatchewan Child and Family Poverty Report 2021. 2021. Available online: https://campaign2000.ca/wp-content/uploads/2021/11/Saskatchewan-Report-Card-English-CPR-2021.pdf (accessed on 2 January 2025).
33. Saskatchewan Teachers' Federation. 2021–2022 Safe Schools Plan. Available online: https://www.stf.sk.ca/resource/2021-22-safe-schools-plan/ (accessed on 4 November 2023).

34. Iachan, R.; Berman, L.; Kyle, T.M.; Martin, K.J.; Deng, Y.; Moyse, D.N.; Middleton, D.; Atienza, A.A. Weighting Nonprobability and Probability Sample Surveys in Describing Cancer Catchment Areas. *Cancer Epidemiol. Biomark. Prev.* **2019**, *28*, 471–477. [CrossRef]
35. Nikolaidis, A.; Paksarian, D.; Alexander, L.; Derosa, J.; Dunn, J.; Nielson, D.M.; Droney, I.; Kang, M.; Douka, I.; Bromet, E.; et al. The Coronavirus Health and Impact Survey (CRISIS) Reveals Reproducible Correlates of Pandemic-Related Mood States across the Atlantic. *Sci. Rep.* **2021**, *11*, 8139. [CrossRef]
36. Kaiser, H.F. An Index of Factorial Simplicity. *Psychometrika* **1974**, *39*, 31–36. [CrossRef]
37. Government of Canada, Public Services and Procurement Canada. Strategies for Population Health: Investing in the Health of Canadians/Prepared by the Federal, Provincial and Territorial Advisory Committee on Population Health: H39-316/1994E-PDF—Government of Canada Publications—Canada.ca. Available online: https://publications.gc.ca/site/eng/9.825914/publication.html (accessed on 10 October 2023).
38. StataCorp. *Stata Statistical Software: Release 17*; StataCorp LLC: College Station, TX, USA, 2021; Stata/SE, 2021.
39. Bouhlila, D.S.; Sellaouti, F. Multiple Imputation Using Chained Equations for Missing Data in TIMSS: A Case Study. *Large-Scale Assess. Educ.* **2013**, *1*, 4. [CrossRef]
40. Baron, R.M.; Kenny, D.A. The Moderator-Mediator Variable Distinction in Social Psychological Research: Conceptual, Strategic, and Statistical Considerations. *J. Personal. Soc. Psychol.* **1986**, *51*, 1173. [CrossRef] [PubMed]
41. Orgilés, M.; Morales, A.; Delvecchio, E.; Francisco, R.; Mazzeschi, C.; Pedro, M.; Espada, J.P. Coping Behaviors and Psychological Disturbances in Youth Affected by the COVID-19 Health Crisis. *Front. Psychol.* **2021**, *12*, 565657. [CrossRef]
42. Brackel, C.L.H.; Lap, C.R.; Buddingh, E.P.; van Houten, M.A.; van der Sande, L.J.T.M.; Langereis, E.J.; Bannier, M.A.G.E.; Pijnenburg, M.W.H.; Hashimoto, S.; Terheggen-Lagro, S.W.J. Pediatric Long-COVID: An Overlooked Phenomenon? *Pediatr. Pulmonol.* **2021**, *56*, 2495–2502. [CrossRef]
43. Buonsenso, D.; Munblit, D.; De Rose, C.; Sinatti, D.; Ricchiuto, A.; Carfi, A.; Valentini, P. Preliminary Evidence on Long COVID in Children. *Acta Paediatr.* **2021**, *110*, 2208–2211. [CrossRef]
44. Szwarcwald, C.L.; Malta, D.C.; Barros, M.B.d.A.; de Souza Júnior, P.R.B.; Romero, D.; de Almeida, W.d.S.; Damacena, G.N.; Werneck, A.O.; da Silva, D.R.P.; Lima, M.G.; et al. Associations of Sociodemographic Factors and Health Behaviors with the Emotional Well-Being of Adolescents during the COVID-19 Pandemic in Brazil. *Int. J. Environ. Res. Public Health* **2021**, *18*, 6160. [CrossRef]
45. Schubert, K.O.; Clark, S.R.; Van, L.K.; Collinson, J.L.; Baune, B.T. Depressive Symptom Trajectories in Late Adolescence and Early Adulthood: A Systematic Review. *Aust. N. Z. J. Psychiatry* **2017**, *51*, 477–499. [CrossRef]
46. Magson, N.R.; Freeman, J.Y.A.; Rapee, R.M.; Richardson, C.E.; Oar, E.L.; Fardouly, J. Risk and Protective Factors for Prospective Changes in Adolescent Mental Health during the COVID-19 Pandemic. *J. Youth Adolesc.* **2021**, *50*, 44–57. [CrossRef]
47. Duan, L.; Shao, X.; Wang, Y.; Huang, Y.; Miao, J.; Yang, X.; Zhu, G. An Investigation of Mental Health Status of Children and Adolescents in China during the Outbreak of COVID-19. *J. Affect. Disord.* **2020**, *275*, 112–118. [CrossRef]
48. Cao, W.; Fang, Z.; Hou, G.; Han, M.; Xu, X.; Dong, J.; Zheng, J. The Psychological Impact of the COVID-19 Epidemic on College Students in China. *Psychiatry Res.* **2020**, *287*, 112934. [CrossRef]
49. Oberle, E.; Ji, X.R.; Kerai, S.; Guhn, M.; Schonert-Reichl, K.A.; Gadermann, A.M. Screen Time and Extracurricular Activities as Risk and Protective Factors for Mental Health in Adolescence: A Population-Level Study. *Prev. Med.* **2020**, *141*, 106291. [CrossRef] [PubMed]
50. LaForge-MacKenzie, K.; Tombeau Cost, K.; Tsujimoto, K.C.; Crosbie, J.; Charach, A.; Anagnostou, E.; Birken, C.S.; Monga, S.; Kelley, E.; Burton, C.L.; et al. Participating in Extracurricular Activities and School Sports during the COVID-19 Pandemic: Associations with Child and Youth Mental Health. *Front. Sports Act. Living* **2022**, *4*, 936041. [CrossRef] [PubMed]
51. Burke, T.A.; Kutok, E.R.; Dunsiger, S.; Nugent, N.R.; Patena, J.V.; Riese, A.; Ranney, M.L. A National Snapshot of U.S. Adolescents' Mental Health and Changing Technology Use during COVID-19. *Gen. Hosp. Psychiatry* **2021**, *71*, 147–148. [CrossRef] [PubMed]
52. Li, X.; Vanderloo, L.M.; Keown-Stoneman, C.D.G.; Cost, K.T.; Charach, A.; Maguire, J.L.; Monga, S.; Crosbie, J.; Burton, C.; Anagnostou, E.; et al. Screen Use and Mental Health Symptoms in Canadian Children and Youth During the COVID-19 Pandemic. *JAMA Netw. Open* **2021**, *4*, e2140875. [CrossRef]
53. Prime, H.; Wade, M.; Browne, D.T. Risk and Resilience in Family Well-Being during the COVID-19 Pandemic. *Am. Psychol.* **2020**, *75*, 631–643. [CrossRef]
54. Wang, C.; Cheong, Y.; Zhu, Q.; Havewala, M.; Ye, Y. Parent Work–Life Conflict and Adolescent Adjustment during COVID-19: Mental Health and Parenting as Mediators. *J. Fam. Psychol.* **2022**, *36*, 325–336. [CrossRef]
55. McArthur, B.A.; Racine, N.; McDonald, S.; Tough, S.; Madigan, S. Child and Family Factors Associated with Child Mental Health and Well-Being during COVID-19. *Eur. Child Adolesc. Psychiatry* **2023**, *32*, 223–233. [CrossRef]
56. Ezpeleta, L.; Navarro, J.B.; de la Osa, N.; Trepat, E.; Penelo, E. Life Conditions during COVID-19 Lockdown and Mental Health in Spanish Adolescents. *Int. J. Environ. Res. Public Health* **2020**, *17*, 7327. [CrossRef]

57. Zhang, X. Household Chaos and Caregivers' and Young Children's Mental Health during the COVID-19 Pandemic: A Mediation Model. *J. Child Fam. Stud.* **2022**, *31*, 1547–1557. [CrossRef]
58. Vallejo Slocker, L.; Sanz Fernández, J.; García Vera, M.P.; Fresneda Sánchez, F.J.; Vallejo Pareja, M.Á. Mental Health, Quality of Life and Coping Strategies in Vulnerable Children during the COVID-19 Pandemic. *Psicothema* **2022**, *34*, 249–258. [CrossRef]
59. Compas, B.E.; Jaser, S.S.; Bettis, A.H.; Watson, K.H.; Gruhn, M.; Dunbar, J.P.; Williams, E.; Thigpen, J.C. Coping, Emotion Regulation and Psychopathology in Childhood and Adolescence: A Meta-Analysis and Narrative Review. *Psychol. Bull.* **2017**, *143*, 939–991. [CrossRef] [PubMed]
60. Ruvalcaba, N.A.; Gallegos, J.; Borges, A.; Gonzalez, N. Extracurricular Activities and Group Belonging as a Protective Factor in Adolescence. *Psicol. Educ.* **2017**, *23*, 45–51. [CrossRef]
61. Heaslip, G.P.; Davis, H.; Barber, B.L. Extracurricular Activity Participation in Early Adolescence Predicts Coping Efficacy One Year Later. *Aust. J. Psychol.* **2021**, *73*, 306–315. [CrossRef]

children

MDPI

Article

Understanding Adolescent Mental Health in the COVID-19 Era: A Psychodynamic Approach

Aslı Akın [1,2] and Lea Sarrar [1,*]

1 Department of Psychology, Faculty of Sciences, MSB Medical School Berlin, 14197 Berlin, Germany
2 Department of Psychiatry and Psychotherapy, Charité-Universitätsmedizin Berlin, 10117 Berlin, Germany; asli.akin@charite.de
* Correspondence: lea.sarrar@medicalschool-berlin.de

Abstract: Objective: This cross-sectional, case-control study aimed to explore the psychodynamic characteristics that influenced adolescents' mental health during the COVID-19 pandemic. Methods: Personality structure impairments, psychodynamic conflicts, defense styles, and mental health issues were examined using the OPD-Structure- and Conflict-Questionnaires, the Defense Style Questionnaire, and the Patient Health Questionnaire in adolescents before (n = 288) and after (n = 451) the COVID-19 pandemic in Germany. Results: Adolescents with mental health issues exhibited greater impairments in personality structure, more immature defense styles, and higher levels of psychodynamic conflicts both before and after the pandemic onset. Comparisons between pre-pandemic and pandemic samples indicated a lower level of the conflict of taking care of oneself versus being cared for in passive mode among adolescents during the pandemic. In-depth analysis of adolescents with mental health issues from the pandemic group revealed significant associations between personality structure impairments and a maladaptive defense style with somatoform, depressive, anxiety, eating, and alcohol use disorders. Conclusion: The findings provide clinically relevant insights into the psychodynamic factors that contributed to the psychological vulnerability of adolescents during the COVID-19 pandemic. These insights can guide the development of targeted psychodynamic interventions to support adolescents' mental health in similar future crises.

Keywords: COVID-19 pandemic; adolescence; mental health; personality structure; psychodynamic conflicts; defense styles

Citation: Akın, A.; Sarrar, L. Understanding Adolescent Mental Health in the COVID-19 Era: A Psychodynamic Approach. *Children* **2024**, *11*, 772. https://doi.org/10.3390/children11070772

Academic Editor: Matteo Alessio Chiappedi

Received: 3 June 2024
Revised: 22 June 2024
Accepted: 24 June 2024
Published: 26 June 2024

1. Introduction

Over the three-year span of the coronavirus disease 2019 (COVID-19) pandemic, the physical and mental health of the global population was at risk [1]. Alongside severe cases of infection and mortality, there was a marked increase in the prevalence of mental health issues following the pandemic's onset [2,3]. Of particular concern was the impact on adolescents, who faced unique challenges during this period of uncertainty and upheaval [4–6]. Adolescents, already navigating a sensitive period of biological, social, cognitive, and emotional transition [7], were further affected by the imposition of social and educational restrictions, as well as ethical responsibilities and challenges, including quarantine of contacts, lockdowns, and vaccinations [8,9].

This global crisis challenged the mental health resilience of adolescents, revealing crucial insights into the psychological factors that contribute to the onset of mental health issues. Understanding these factors requires a comprehensive theoretical approach that can account for the complexities of adolescent psychology. Psychodynamic theory provides a clinically and empirically established framework for exploring these psychological factors that increase psychological vulnerability in adolescents, focusing on three core concepts: impairments in personality structure, psychodynamic conflicts, and defense mechanisms [10].

Personality structure, which mainly develops during the earliest attachment and relationship experiences, includes basic mental functions that regulate the self and its relationships with others [11]. With its sub-domains of attachment, identity, interpersonality, and control, the personality structure can achieve varying levels of structural functioning depending on the extent of impairments [10]. Low levels of structural functioning are linked to general psychopathology [12], making adolescents more susceptible to mental health issues during stressful times like a pandemic. Adolescents with impairments in personality structure struggle more with emotional regulation and interpersonal relationships [11], which could exacerbate the psychological impact of the pandemic.

Psychodynamic conflicts are defined as unconscious, temporally persistent, opposing motives, desires, values, and ideas within a person [13]. They arise from developmental tasks every child goes through and can concern the following conflictual needs and motives: closeness versus distance, submission versus control, taking care of oneself versus being cared for, self-worth conflict, guilt conflict, oedipal conflict, and identity conflict [10]. Psychodynamic conflicts shape experience and actions through either an active or passive mode of coping with them; while the passive mode involves unfulfilled desires like the need for security, the active mode involves defenses against these deficits, such as striving for self-sufficiency and suppressing attachment wishes [10]. According to the fundamental psychoanalytic theory, repressed psychodynamic conflicts that originate in childhood can be triggered by stressful events later in life and lead to the formation of symptoms [14], making adolescents more prone to anxiety, depression, and other mental health problems during the pandemic crisis.

Defense mechanisms are unconscious psychological strategies that help ward off unpleasant feelings [10,15]. These mechanisms become active when psychodynamic conflicts or structural impairments trigger too much emotional tension, preventing these feelings from becoming conscious [16]. While adaptive defense mechanisms ensure healthy and functional ways of dealing with increased emotional tension, neurotic and particularly maladaptive defense mechanisms are associated with psychopathology [15,17]. Adolescents who rely on maladaptive defense mechanisms may have experienced heightened psychological distress during the pandemic.

Overall, structural functioning, psychodynamic conflicts, and defense mechanisms represent psychological vulnerability factors that mainly develop throughout childhood and have a decisive influence on the ability to deal with stressful situations in adolescence and later adulthood [10]. The COVID-19 pandemic was a particularly stressful situation, or even a "global trauma" [18], that challenged people's structural functioning, triggered psychodynamic conflicts, and required the use of defense mechanisms. Studies on adults have already confirmed the importance of psychodynamic characteristics for mental health during the pandemic, showing that defense mechanisms mediate the relationship between personality traits and post-traumatic symptoms [19] and predict adherence to COVID-19-related conspiracy theories [20]. A recently published study also demonstrated the positive relationship between impairments in personality functioning and psychological distress in adults during the pandemic [21].

Despite these findings, the specific psychodynamic vulnerability factors of adolescents during the pandemic remain underexplored. This study aims to elucidate the relevance of impairments in personality structure, psychodynamic conflicts, and defense mechanisms for the mental health status of adolescents during the COVID-19 pandemic. Specifically, we examined the psychodynamic characteristics of adolescents from clinical and non-clinical samples before and after the outbreak of the pandemic in Germany. In addition to a descriptive comparison of psychodynamic characteristics between adolescents with and without mental health issues, as well as before and during the pandemic, we investigated whether and to what extent various forms of mental health issues in adolescents during the COVID-19 pandemic can be explained by their psychodynamic characteristics. In sum, this approach aims to deepen our understanding of the complex psychodynamic interactions that shape adolescents' mental health in times of crisis, enabling the development of targeted interventions to strengthen their resilience.

2. Materials and Methods

2.1. Sample

A total of 739 adolescents, aged between 14 and 21 years (M_{age} = 17.6; *SD* = 2.1; 61% female, 38% male, and 1% non-binary) took part in the study. Participants were recruited from outpatient clinics and educational institutions across Germany. Of the total sample, 39% (*n* = 288, M_{age} = 17.2; *SD* = 1.8; 64% female, 35% male, and 1% non-binary) were recruited and assessed before the outbreak of the COVID-19 pandemic in Germany (between March 2019 and February 2020). The remaining 61% (*n* = 451, M_{age} = 17.8; *SD* = 2.2; 58% female, 40% male, and 2% non-binary) were surveyed during the pandemic period (between March 2020 and March 2023). Among the 288 adolescents surveyed before the outbreak of the pandemic, 28% (*n* = 81) met the criteria for at least one mental disorder syndrome according to the Patient Health Questionnaire (PHQ-D) [22], while 72% (*n* = 207) showed no signs of mental health issues. Conversely, of the 451 adolescents surveyed during the pandemic, 43% (*n* = 192) met the criteria for at least one mental disorder syndrome according to the PHQ-D [22], while 57% (*n* = 255) showed no signs of mental health issues. Most participants were from high (41% of the pre-pandemic and 50% of the pandemic sample) or medium (39% of the pre-pandemic and 27% of the pandemic sample) socioeconomic backgrounds. Less than one third of the participants lived under low or very low socioeconomic conditions (20% of the pre-pandemic and 23% of the pandemic sample).

2.2. Measures

All participants were assigned to fill out a set of self-assessment questionnaires to measure their psychodynamic characteristics and mental health status. The questionnaires were administered by mail as paper-pencil versions. In order to prevent context effects, the questionnaires were given in a randomized order.

The OPD-CA2 Structure Questionnaire (OPD-CA2-SQ) [23] was used to assess impairments in personality structure. The questionnaire includes 81 items that assess impairments across four domains of personality structure, using a five-point scale ranging from no (0) to yes (4). Higher scores signify greater impairment in personality structure, indicating lower levels of structural functioning. The overall mean score of all items provides a general measure of personality structure impairments. For our analysis, we computed mean scores based on the raw scale scores, differing from the original authors' scoring guidelines. The overall scale demonstrated very high reliability (McDonald's ω = 0.97), and the sub-domains of attachment (McDonald's ω = 0.86), identity (McDonald's ω = 0.99), interpersonality (McDonald's ω = 0.82), and control (McDonald's ω = 0.90) showed high-to-very-high reliabilities in our sample. The results align with the similarly high internal consistencies reported in other studies [24].

The OPD-CA Conflict Questionnaire (OPD-CA-CQ) [25] was employed to assess the active and passive modes of coping with the seven psychodynamic conflicts. The questionnaire consists of 28 items, rated on a five-point scale from no (0) to yes (4). Higher average scores on the conflict scales indicate more pronounced psychodynamic conflicts. In this sample, the two-item scales turned out to be very reliable for the conflict of closeness versus distance in active (Spearman–Brown ρ = 0.97) and passive mode (Spearman–Brown ρ = 0.96), submission versus control in active (Spearman–Brown ρ = 0.98) and passive mode (Spearman–Brown ρ = 0.96), taking care of oneself versus being cared for in active (Spearman–Brown ρ = 0.98) and passive mode (Spearman–Brown ρ = 0.95), self-worth conflict in active (Spearman–Brown ρ = 0.98) and passive mode (Spearman–Brown ρ = 0.97), guilt conflict in active (Spearman–Brown ρ = 0.99) and passive mode (Spearman–Brown ρ = 0.97), oedipal conflict in active (Spearman–Brown ρ = 0.98) and passive mode (Spearman–Brown ρ = 0.96), and identity conflict in active (Spearman–Brown ρ = 0.97) and passive mode (Spearman–Brown ρ = 0.98). Prior psychometric investigations of the OPD-CA-CQ revealed inconsistent reliabilities for some subscales [26].

The Defense Style Questionnaire for Adolescents (DSQ-40-A) [27] was used to assess adaptive, neurotic, and maladaptive defense styles. This questionnaire comprises 40 items, representing 20 defense mechanisms, each with two items rated on a nine-point scale from not true (0) to completely true (8). Scores for defense mechanisms and defense styles are calculated by averaging the ratings for the relevant items, with higher scores indicating higher severity of the defense mechanisms or styles. The adaptive (McDonald's ω = 0.54) and neurotic (McDonald's ω = 0.50) defense style scales were sufficiently reliable, while the maladaptive defense style scale (McDonald's ω = 0.75) demonstrated acceptable reliability in our sample. Previous psychometric evaluations of the DSQ-40 in adolescents have shown similar internal consistencies [28].

The PHQ-D [22] was utilized to screen for common mental disorders at the syndrome level. The 58-item questionnaire measures 16 conditions across five categories: somatoform, depressive, anxiety, eating, and alcohol use disorders. In this study, the somatoform syndromes scale (McDonald's ω = 0.69) and depressive syndromes scale (McDonald's ω = 0.80) showed acceptable-to-good reliabilities. Previous psychometric analyses of the PHQ-D have indicated excellent internal consistencies [29]. For the remaining diagnostic scales, calculating internal consistencies is deemed unnecessary, as these are primarily evaluated categorically with specific jump rules.

2.3. Data Analyses

Statistical analyses were conducted using IBM SPSS Statistics (Version 29). All tests were two-tailed, with a p-value < 0.05 considered statistically significant. Descriptive statistics, including means and standard deviations, were computed for all psychodynamic variables across each study group.

To compare impairments in personality structure, the use of adaptive, neurotic, and maladaptive defense mechanisms, and the extents of psychodynamic conflicts between adolescents with and without mental health issues as well as between the pre-pandemic and pandemic samples, separate univariate analyses of covariance (ANCOVA) were performed, controlling for sociodemographic variables (age, sex, and socioeconomic conditions). To control the type 1 error rate in multiple comparisons, the Bonferroni–Holm method was applied to all significance values. Partial η^2 values were interpreted as follows: $\geq$0.01 as a small effect, $\geq$0.06 as a moderate effect, and $\geq$0.14 as a large effect [30].

To investigate the importance of psychodynamic characteristics for different types of mental health issues in adolescents during the COVID-19 pandemic, a series of binary logistic regression analyses were performed. Initially, the dependent variable was the general presence of mental health issues (coded as 0 for presence and 1 for absence of any mental health issues). Subsequently, syndromes from the PHQ-D [22] relating to somatoform, depressive, anxiety, eating, and alcohol use disorders were considered as dichotomous outcome variables (coded as 0 for presence and 1 for absence of the syndrome). The offered predictors were impairments in the overall and sub-domains of personality structure, the seven psychodynamic conflicts, and the three defense styles, as well as age, sex (coded as 0 for female and 1 for male sex), and socioeconomic conditions. Due to the exploratory nature of this study, independent variables were introduced into the regression models using a forward stepwise method, i.e., variables with $p \geq 0.05$ were eliminated stepwise. Nagelkerke R^2 values were obtained for all models to estimate their explanatory power, with $\geq$0.20 interpreted as a small effect, $\geq$0.40 as a moderate effect, and $\geq$0.50 as a large effect [31].

3. Results

3.1. Comparisons between Adolescents with and without Mental Health Issues

Descriptive statistics, including means, standard deviations, and group comparisons for impairments in personality structure, defense styles, and psychodynamic conflicts among adolescents with and without mental health issues from the pre-pandemic and pandemic samples, are presented in Table 1.

Table 1. Psychodynamic characteristics of adolescents with and without mental health issues before and during the COVID-19 pandemic.

	Pre-Pandemic Sample (n = 288)				Pandemic Sample (n = 451)			
	Adolescents with Mental Health Issues (n = 81)	Adolescents without Mental Health Issues (n = 207)	Comparison between Groups [a]		Adolescents with Mental Health Issues (n = 192)	Adolescents without Mental Health Issues (n = 255)	Comparison between Groups [a]	
Variable	M (SD)	M (SD)	F(1, 266)	η^2	M (SD)	M (SD)	F(1, 366)	η^2
Overall PS	1.65 (0.54)	1.08 (0.48)	72.59 *	0.22	1.75 (0.70)	1.04 (0.57)	94.19 *	0.21
Attachment	1.55 (0.63)	1.01 (0.50)	55.81 *	0.17	1.70 (0.73)	1.00 (0.54)	91.38 *	0.20
Identity	1.73 (0.59)	1.17 (0.55)	56.69 *	0.18	1.79 (0.76)	1.15 (0.55)	67.37 *	0.16
Interpersonality	1.62 (0.54)	1.07 (0.52)	59.75 *	0.18	1.71 (0.71)	1.03 (0.67)	77.58 *	0.18
Control	1.66 (0.64)	1.03 (0.59)	62.34 *	0.19	1.78 (0.82)	0.93 (1.00)	56.88 *	0.14
Adaptive DS	4.45 (1.23)	4.99 (1.13)	11.33 *	0.04	4.62 (1.38)	4.83 (1.34)	2.13	0.01
Neurotic DS	3.88 (1.05)	3.55 (1.05)	5.83 *	0.02	4.01 (1.13)	3.44 (1.33)	16.84 *	0.04
Maladaptive DS	3.17 (1.09)	2.37 (0.86)	46.81 *	0.15	3.21 (1.04)	2.37 (1.03)	59.61 *	0.14
C1a	1.23 (0.81)	0.93 (0.85)	6.57 *	0.02	0.98 (0.86)	0.26 (6.51)	1.05	0.00
C1p	1.73 (0.95)	1.45 (0.94)	3.57	0.01	1.71 (1.10)	1.01 (6.58)	0.66	0.00
C2a	1.16 (0.72)	1.01 (0.65)	2.18	0.01	1.06 (0.77)	0.60 (6.52)	0.48	0.00
C2p	1.72 (0.75)	1.46 (0.78)	6.98 *	0.03	1.83 (0.82)	1.03 (6.56)	1.22	0.00
C3a	1.05 (0.85)	0.80 (0.73)	6.72 *	0.03	1.32 (0.86)	0.44 (6.52)	1.62	0.00
C3p	1.86 (0.70)	1.91 (0.65)	0.84	0.00	1.34 (0.84)	1.02 (6.57)	0.19	0.00
C4a	1.61 (115)	1.39 (1.01)	2.10	0.01	1.75 (1.17)	1.05 (6.60)	0.87	0.00
C4p	1.48 (0.99)	0.91 (0.84)	21.33 *	0.07	1.53 (1.14)	0.56 (6.54)	1.78	0.01
C5a	0.29 (0.56)	0.13 (0.38)	8.58 *	0.03	0.55 (0.88)	−0.19 (6.46)	1.13	0.00
C5p	1.60 (1.00)	1.46 (0.98)	0.44	0.00	1.76 (1.10)	1.06 (6.59)	0.94	0.00
C6a	0.86 (0.94)	0.67 (0.87)	4.28 *	0.02	0.91 (0.97)	0.38 (6.54)	0.54	0.00
C6p	1.33 (0.86)	1.14 (0.88)	3.04	0.01	1.18 (0.99)	0.71 (6.55)	0.53	0.01
C7a	1.15 (0.86)	0.80 (0.72)	11.50 *	0.04	1.08 (0.98)	0.43 (6.53)	1.09	0.00
C7p	1.44 (1.16)	0.74 (0.89)	28.31 *	0.10	1.42 (1.24)	0.43 (6.54)	1.87	0.01

Note. N = 739. PS = personality structure. DS = defense style. C1a = conflict of closeness versus distance in active mode. C1p = conflict of closeness versus distance in passive mode. C2a = conflict of submission versus control in active mode. C2p = conflict of submission versus control in passive mode. C3a = conflict of taking care of oneself versus being cared for in active mode. C3p = conflict of taking care of oneself versus being cared for in passive mode. C4a = conflict of self-worth in active mode. C4p = conflict of self-worth in passive mode. C5a = guilt conflict in active mode. C5p = guilt conflict in passive mode. C6a = oedipal conflict in active mode. C6p = oedipal conflict in passive mode. C7a = identity conflict in active mode. C7p = identity conflict in passive mode. M = mean. SD = standard deviation. F = test value. η^2 = partial eta squared. [a] Age, sex, and socioeconomic conditions were controlled * Bonferroni–Holm adjusted $p \leq 0.050$.

3.2. Comparisons between Pre-Pandemic and Pandemic Samples

After adjusting for age, sex, and socioeconomic conditions, adolescents with mental health issues from the pre-pandemic and pandemic samples showed significant differences regarding the conflict of taking care of oneself versus being cared for in passive mode, $F(1, 228) = 18.66$, $p_{adj} = 0.003$, partial $\eta^2 = 0.08$. After adjusting for age, sex, and socioeconomic conditions, adolescents without mental health issues from the pre-pandemic and pandemic samples also showed significant differences regarding the conflict of taking care of oneself versus being cared for in passive mode, $F(1, 404) = 3.93$, $p_{adj} = 0.048$, partial $\eta^2 = 0.01$.

3.3. Predictors of Mental Health Issues during the COVID-19 Pandemic

Multiple binary logistic regression models were used to identify the psychodynamic predictors of mental health issues among adolescents during the COVID-19 pandemic. The results are summarized in Table 2.

Table 2. Binary logistic regression models predicting the presence of mental health issues in adolescents during the COVID-19 pandemic based on psychodynamic and demographic characteristics.

Variable	B	SE_B	OR	95% CI	p
Model for predicting general mental health issues (n = 192)					
impairments in the overall personality structure	2.92	0.58	0.05	0.02, 0.17	<0.001
impairments in the personality structure domain of identity	1.13	0.52	3.09	1.12, 8.55	0.030
age	0.24	0.06	0.79	0.70, 0.88	<0.001
Model for predicting somatoform syndrome (n = 58)					
impairments in the overall personality structure	1.40	0.26	4.03	2.43, 6.69	<0.001
sex	−1.80	0.55	0.17	0.06, 0.49	0.001
Model for predicting major depressive syndrome (n = 66)					
impairments in the overall personality structure	2.55	0.41	12.76	5.71, 28.53	<0.001
maladaptive defense style	0.50	0.23	1.64	1.06, 2.55	0.028
Model for predicting other depressive syndrome (n = 31)					
impairments in the personality structure domain of interpersonality	0.63	0.28	1.88	1.08, 3.26	0.025
Model for predicting panic syndrome (n = 28)					
impairments in the personality structure domain of attachment	1.48	0.35	4.40	2.23, 8.67	<0.001
Model for predicting other anxiety syndrome (n = 31)					
impairments in the personality structure domain of control	1.69	0.55	5.40	1.83, 15.95	0.002
impairments in the personality structure domain of attachment	1.15	0.57	3.17	1.05, 9.60	0.041
age	0.32	0.14	1.38	1.05, 1.81	0.022
Model for predicting syndrome of eating disorder (n = 19)					
impairments in the personality structure domain of identity	1.48	0.35	4.38	2.19, 8.76	<0.001
Model for predicting harmful alcohol consumption (n = 131)					
maladaptive defense style	0.39	0.11	1.48	1.18, 1.84	<0.001
sex	0.60	0.24	1.83	1.14, 2.94	0.012
age	0.13	0.06	1.13	1.02, 1.27	0.026

Note. N = 192. B = regression coefficient. SE_B = standard error of B. OR = odds ratio. CI = confidence interval.

The model predicting the general presence of mental health issues was statistically significant, $\chi^2(3) = 111.35$, $p < 0.001$, and explained a small amount of variance, Nagelkerke's $R^2 = 0.35$. Key predictors included impairments in overall personality structure, impairments in the personality structure domain of identity, and age.

The model predicting the presence of a somatoform syndrome was statistically significant, $\chi^2(2) = 68.18$, $p < 0.001$, explaining a small amount of variance, Nagelkerke's $R^2 = 0.31$. Impairments in the overall personality structure and age were significant predictors.

The model predicting the presence of a major depressive syndrome was statistically significant, $\chi^2(2) = 128.61$, $p < 0.001$, and explained a large amount of variance, Nagelkerke's $R^2 = 0.51$. Significant predictors included maladaptive defense style and impairments in overall personality structure.

The model predicting the presence of another depressive syndrome with impairments in the personality structure domain of interpersonality as the best predictor was statistically significant, $\chi^2(1) = 5.00$, $p = 0.025$, explaining a small amount of variance, Nagelkerke's $R^2 = 0.04$.

The model predicting the presence of a panic syndrome with impairments in the personality structure domain of attachment as the best predictor was statistically significant, $\chi^2(2) = 39.84$, $p < 0.001$, explaining a small amount of variance, Nagelkerke's $R^2 = 0.30$.

The model predicting the presence of another anxiety syndrome was statistically significant, $\chi^2(3) = 67.48$, $p < 0.001$, explaining a moderate amount of variance, Nagelkerke's $R^2 = 0.46$. Significant predictors included impairments in the personality structure domains of control and attachment, as well as age.

The model predicting the presence of an eating disorder syndrome with impairments in the personality structure domain of identity as the best predictor was statistically significant, $\chi^2(1) = 18.89$, $p < 0.001$, explaining a small amount of variance, Nagelkerke's $R^2 = 0.17$.

The model predicting the presence of harmful alcohol consumption was statistically significant, $\chi^2(3) = 26.42$, $p < 0.001$, explaining a small amount of variance, Nagelkerke's $R^2 = 0.10$. Significant predictors included maladaptive defense style, sex, and age.

4. Discussion

The present study aimed to explore the relevance of adolescents' psychodynamic characteristics in relation to their mental health during the COVID-19 pandemic.

First of all, by comparing the mental health status of adolescents before and after the outbreak of the COVID-19 pandemic, evidence for the increased prevalence of mental health problems in adolescents after the outbreak of the pandemic could be provided. According to the observed cross-sectional data, the percentage of adolescents experiencing mental health issues increased from 28% to 43%, i.e., by a total of 15% after the outbreak of the pandemic. The discovered prevalence of mental health issues during the COVID-19 pandemic aligns with other studies [32–34], supporting the generalizability of the study results discussed below.

In accordance with psychodynamic theory, the findings of the study confirmed that adolescents with mental health issues exhibited higher impairments in personality structure, more frequent use of maladaptive and neurotic defense mechanisms, and higher extents of psychodynamic conflicts compared to their peers without mental health issues, both before and after the outbreak of the pandemic. In particular, the large mean differences between the clinical and non-clinical subgroups with regard to impairments in the overall and sub-domains of personality structure illustrate the relevance of the basic mental functions developed in the earliest years of life for general mental health [10,11]. There were also large mean differences between adolescents with and without mental health issues in terms of maladaptive defense style. Since defense mechanisms are understood as part of the personality structure and as a qualitative criterion for assessing it [10], this further underscores the high impact of psychodynamic characteristics developed early in life on mental health. In summary, impairments in personality structure and the use of maladaptive defense mechanisms appear to be general risk factors for the development of mental health issues in adolescents.

After applying the Bonferroni–Holm correction, we found almost no significant differences between adolescents from the pre-pandemic and pandemic samples regarding the psychodynamic characteristics. Interestingly, the only significant difference was found for the conflict of taking care of oneself versus being cared for in passive mode. This conflict was significantly higher in both adolescents with and without mental health issues before the pandemic, so it can be assumed that the outbreak of the pandemic decreased the extent of the conflict of taking care of oneself versus being cared for in passive mode. The central motive of this conflict is dependence in relationships and the need to receive attention, security, and care [10]. In the passive mode of coping with this conflict, demanding and clinging relationship behavior occurs and the fear of not being adequately cared for is constantly present [10]. Against the background of the consequences of the COVID-19 lockdown, such as stay-at-home-orders, school closures, and working from home, which led parents and their children to be closer together, it can be assumed that adolescents experienced increased care and attention in the parental home and thus could partially resolve this conflict. Adolescents, who are particularly confronted with the desire for independence and the simultaneous need for connection as part of the process of natural separation from their parents, may have experienced in the context of the COVID-19 pandemic that, if insecurity and danger arise in the outside world, they can find security and care in the parental home. The potential for such beneficial effects of the pandemic on family coexistence has already been demonstrated in a qualitative study by Subhadra et al. [35].

In turn, the fact that the extent of the impairments in personality structure and the defense styles hardly differ between the pre-pandemic and pandemic samples can be understood in light of the profound psychoanalytic theory of personality development. The psychoanalytic theory suggests that structural functioning and defense mechanisms are temporally and situationally stable unconscious aspects of personality—or rather *neurotic* dispositions [36]. However, these stable aspects of personality can become problematic and lead to mental illness when an individual faces particularly stressful life events.

Ultimately, the binary logistic regression analyses provided further insights into the associations between psychodynamic characteristics and various mental health syndromes in adolescents during the pandemic.

The regression models for predicting the presence of general and specific mental health issues indicated that impairments in personality structure and the use of maladaptive defense mechanisms are associated with a broad range of psychopathology during the pandemic. Accordingly, previous empirical studies that demonstrated the importance of personality structure and defense mechanisms for the mental health of adults during the COVID-19 pandemic could now be confirmed in adolescents [19–21].

In addition, the findings indicated that especially impairments in the personality structure sub-domain of identity appear to be a relevant risk factor for the development of mental health issues in adolescents during the pandemic. Khazand et al. [37] have already reported that the pandemic-related changes in adolescents' social interaction opportunities—e.g., through closures of schools, distance learning, or the cancellation of free time activities—made the processes of exploring social roles and selecting peer groups, which are crucial for identity development, more difficult. The link between impairments in the sub-domain of identity and syndromes of eating disorders is in turn consistent with the tendency of adolescents, described by Klotter [38], to process identity problems through their bodies when other opportunities to experience self-determination are unavailable.

Another key finding of the regression analysis was the positive predictive effect of a maladaptive defense style on syndromes of major depression, eating disorders, and harmful alcohol consumption among adolescents during the pandemic. Maladaptive defense mechanisms, such as denial, splitting, and projection, which refuse the acceptance of reality and lead to a view of things as either all-good or all-bad [15], appear to have made it particularly difficult for adolescents to deal with the pandemic and contributed to various mental health problems. By additionally taking into account the significantly increased extent of the adaptive defense style in adolescents without mental health issues, the current investigation demonstrated the importance of defense mechanisms for the mental health state of adolescents during the COVID-19 pandemic.

Moreover, the regression analysis results revealed that panic and other anxiety syndromes in adolescents during the COVID-19 pandemic are primarily linked to impairments in the personality structure domain of attachment. This finding aligns with previous studies, which have demonstrated that insecure attachment styles predict COVID-19-related anxiety symptoms [39] and that attachment security has helped adolescents manage psychological stress and influenced their prosocial and health-protective behaviors during the pandemic [40].

In addition to the predictive relevance of psychodynamic characteristics for various syndromes of mental health issues, gender- and age-specific associations could be discovered.

Adolescents' sex was negatively related to the presence of a somatoform syndrome and positively related to harmful alcohol consumption. Because of the small number of participants, adolescents with non-binary genders were excluded from the regression analyses, and female and male sex were considered as dichotomous variables. Accordingly, the results imply that female adolescents were more affected by a somatoform syndrome and male adolescents by harmful alcohol consumption during the COVID-19 pandemic. Previous studies have also pointed out the importance of gender-specific examinations of mental health problems during the pandemic [41,42].

Adolescents' age was positively associated with the presence of general mental health issues, other anxiety syndromes, and harmful alcohol consumption, indicating that, the older adolescents were during the COVID-19 pandemic, the higher their risk of mental health problems was. This finding is in line with previous studies, which reported higher levels of mental health symptoms in older adolescents than in younger during the pandemic [43,44], but also with the discovery that older adolescents are at increased risk of mental disorders outside of the context of the pandemic [45].

Ultimately, it can be said that the present study had notable strengths, including a large sample of adolescents and control groups from the pre-pandemic period and adolescents without mental health issues, but also limitations which should be taken into account when interpreting the results. An important limitation is the cross-sectional design, which hinders causal interpretations of the relationships between psychodynamic characteristics developed early in life and the onset of mental health issues during the COVID-19 pandemic. Moreover, the study sample, while diverse, may not be fully representative of the broader adolescent population. The majority of participants were from medium to high socioeconomic backgrounds, which could limit the generalizability of the findings to adolescents from lower socioeconomic conditions. Additionally, it is important to consider the unequal distribution of adolescents with mental health issues within the subsamples, as well as the lack of consideration for clinically existing comorbidities. Furthermore, the use of self-report questionnaires may introduce bias due to subjective interpretation and response tendencies [46]. The method of participant recruitment through outpatient clinical facilities and educational institutions may have also introduced selection bias, as it might not capture adolescents who did not seek mental health support or were not enrolled in educational institutions during the pandemic. Future studies should employ longitudinal designs and incorporate multi-method approaches to provide a more comprehensive understanding of the psychodynamic factors influencing youth mental health during and beyond the pandemic.

5. Conclusions

The present cross-sectional, case-control study provides empirical evidence highlighting the importance of psychodynamic characteristics for the mental health status of adolescents during the COVID-19 pandemic. The findings demonstrate that syndromes of somatoform, depressive, anxiety, eating, and alcohol use disorders in adolescents during the pandemic were linked to impairments in overall and sub-domains of personality structure, as well as to a maladaptive defense style. Psychodynamic-oriented preventive and therapeutic interventions targeting these specific vulnerabilities might be effective in mitigating the impact of the pandemic on youth mental health. However, the conclusions should be interpreted with caution given the study's limitations. Future research exploring these psychodynamic characteristics through longitudinal designs and incorporating a more diverse sample is needed to enhance our understanding and treatment of youth mental health in the context of unprecedented societal stressors.

Author Contributions: Conceptualization and methodology, A.A. and L.S.; investigation, resources, and data curation, L.S.; formal analysis, writing—original draft preparation, A.A.; writing—review and editing, L.S.; supervision, L.S. All authors have read and agreed to the published version of the manuscript.

Funding: The article processing charge (APC) for this publication was funded by the MSB Medical School Berlin.

Institutional Review Board Statement: This study was conducted in compliance with the Declaration of Helsinki and was approved by the Research Ethics Board at MSB Medical School Berlin (approval number: MSB-2020/30; date of approval: 30 March 2020).

Informed Consent Statement: All adolescents involved were informed about the study's data usage and protection measures, and provided written informed consent. Written informed consent was also obtained from legal guardians for participants under the age of 16.

Data Availability Statement: The datasets generated and analyzed during the current study are not publicly available due to confidentiality reasons. Requests to access the datasets should be addressed to L.S.

Conflicts of Interest: The authors declare that they have no conflicts of interest.

References

1. Onyeaka, H.; Anumudu, C.K.; Al-Sharify, Z.T.; Egele-Godswill, E.; Mbaegbu, P. COVID-19 pandemic: A review of the global lockdown and its far-reaching effects. *Sci. Prog.* **2021**, *104*, 19854. [CrossRef]
2. Nochaiwong, S.; Ruengorn, C.; Thavorn, K.; Hutton, B.; Awiphan, R.; Phosuya, C.; Ruanta, Y.; Wongpakaran, N.; Wongpakaran, T. Global prevalence of mental health issues among the general population during the coronavirus disease-2019 pandemic: A systematic review and meta-analysis. *Sci. Rep.* **2021**, *11*, 10173. [CrossRef]
3. Clemente-Suárez, V.J.; Martínez-González, M.B.; Benitez-Agudelo, J.C.; Navarro-Jiménez, E.; Beltran-Velasco, A.I.; Ruisoto, P.; Diaz Arroyo, E.; Laborde-Cárdenas, C.C.; Tornero-Aguilera, J.F. The Impact of the COVID-19 Pandemic on Mental Disorders. A Critical Review. *Int. J. Environ. Res. Public Health* **2021**, *18*, 10041. [CrossRef]
4. Jones, E.A.K.; Mitra, A.K.; Bhuiyan, A.R. Impact of COVID-19 on Mental Health in Adolescents: A Systematic Review. *Int. J. Environ. Res. Public Health* **2021**, *18*, 2470. [CrossRef]
5. Branje, S. The impact of the COVID-19 pandemic on adolescent mental health across the world. *Curr. Opin. Psychol.* **2023**, *53*, 101665. [CrossRef]
6. Meherali, S.; Punjani, N.; Louie-Poon, S.; Abdul Rahim, K.; Das, J.K.; Salam, R.A.; Lassi, Z.S. Mental health of children and adolescents amidst COVID-19 and past pandemics: A rapid systematic review. *Int. J. Environ. Res. Public Health* **2021**, *18*, 3432. [CrossRef]
7. Kroger, J. *Identity in Adolescence*; Routledge: London, UK, 2004.
8. Longobardi, C.; Morese, R.; Fabris, M.A. COVID-19 emergency: Social distancing and social exclusion as risks for suicide ideation and attempts in adolescents. *Front. Psychol.* **2020**, *11*, 551113. [CrossRef]
9. Marrone, M.; Luca, B.P.D.; Stellacci, A.; Buongiorno, L.; Caricato, P.; Cazzato, G.; Ferorelli, D.; Solarino, B.; Stefanizzi, P.; Tafuri, S.; et al. COVID-19 Vaccination in Italian Children: The Limits of Parental Rights. *Children* **2022**, *9*, 625. [CrossRef]
10. Task Force OPD-CA-2. (Ed.) *OPD-CA-2 Operationalisierte Psychodynamische Diagnostik im Kindes- und Jugendalter: Grundlagen und Manual [OPD-CA-2 Operationalized Psychodynamic Diagnosis in Childhood and Adolescence: Theoretical Basis and User Manual]*; Hogrefe: Bern, Switzerland, 2020.
11. Resch, F.; Parzer, P. (Eds.) Psychic Structure, personality and psychodynamics. In *Adolescent Risk Behavior and Self-Regulation: A Cybernetic Perspective*; Springer: Cham, Switzerland, 2021; pp. 55–63.
12. Bach, B.; Simonsen, S. How does level of personality functioning inform clinical management and treatment? Implications for ICD-11 classification of personality disorder severity. *Curr. Opin. Psychiatry* **2021**, *34*, 54–63. [CrossRef]
13. Simmonds, J.; Constantinides, P.; Perry, J.C.; Drapeau, M.; Sheptycki, A.R. Assessing psychodynamic conflict. *Psychodyn. Psychiatry* **2015**, *43*, 349–377. [CrossRef]
14. Butler, S.F.; Demmin, H.; Strupp, H.H. Psychodynamic psychotherapy. In *Introduction to Clinical Psychology*; Heiden, L.A., Hersen, M., Eds.; Springer: Boston, MA, USA, 1995; pp. 213–229.
15. Cramer, P. Defense mechanisms: 40 years of empirical research. *J. Personal. Assess.* **2015**, *97*, 114–122. [CrossRef]
16. Heinemann, E.; Hopf, H. *Psychische Störungen in Kindheit und Jugend [Mental Disorders in Childhood and Adolescence]*; Kohlhammer: Stuttgart, Germany, 2021.
17. Strandholm, T.; Kiviruusu, O.; Karlsson, L.; Miettunen, J.; Marttunen, M. Defense Mechanisms in Adolescence as Predictors of Adult Personality Disorders. *J. Nerv. Ment. Dis.* **2016**, *204*, 349–354. [CrossRef]
18. Viscuso, D.G.I.; Mangiapane, D.E. Pandemic COVID-19: Psychodynamic analysis of a global trauma. Clinical considerations pre\post Lock down. *J. Med. Res. Health Sci.* **2020**, *3*, 976–990. [CrossRef]
19. Gori, A.; Topino, E.; Palazzeschi, L.; Di Fabio, A. Which personality traits can mitigate the impact of the pandemic? Assessment of the relationship between personality traits and traumatic events in the COVID-19 pandemic as mediated by defense mechanisms. *PLoS ONE* **2021**, *16*, e0251984. [CrossRef]
20. Gioia, F.; Imperato, C.; Boursier, V.; Franceschini, C.; Schimmenti, A.; Musetti, A. The role of defense styles and psychopathological symptoms on adherence to conspiracy theories during the COVID-19 pandemic. *Sci. Rep.* **2023**, *13*, 3482. [CrossRef]
21. Kampe, L.; Hörz-Sagstetter, S.; Bohn, J.; Remmers, C. How personality functioning relates to psychological distress and behavioral attitudes during the COVID-19 pandemic. *Eur. Arch. Psychiatry Clin. Neurosci.* **2024**. [CrossRef]
22. Löwe, B.; Spitzer, R.L.; Zipfel, S.; Herzog, W. *Gesundheitsfragebogen für Patienten (PHQ-D). Manual und Testunterlagen [Patient Health Questionnaire (PHQ-D). Manual and Test Documents]*; Pfizer: Karlsruhe, Germany, 2002.

23. Goth, K.; Schrobildgen, C.; Schmeck, K. Das Inventar OPD-KJ2-SF (Operationalisierte Psychodynamische Diagnostik im Kindes- und Jugendalter—Strukturfragebogen). Deutschsprachige Version: Ein Fragebogen zur Selbstbeantwortung für die Erfassung der Persönlichkeitsstruktur im Jugendalter—Kurzbeschreibung. [The Inventory OPD-CA2-SQ (Operationalized Psychodynamic Diagnostics in Childhood and Adolescence—Structural Questionnaire). German Version: A Questionnaire for Self-Response for the Assessment of Personality Structure in Adolescence—Short Description]. Academic-Tests. 2018. Available online: https://academic-tests.com (accessed on 1 May 2019).
24. Schrobildgen, C.; Goth, K.; Weissensteiner, R.; Lazari, O.; Schmeck, K. Der OPD-KJ2-SF—Ein Instrument zur Erfassung der Achse Struktur der OPD-KJ-2 bei Jugendlichen im Selbsturteil [The OPD-CA2-SQ—An instrument for assessing the structure axis of the OPD-CA-2 in adolescents' self-assessment]. *Z. Kinder- Jugendpsychiatrie Psychother.* **2019**, *47*, 428–440. [CrossRef]
25. Seiffge-Krenke, I.; Escher, F.J. Entwicklung eines Fragebogens zur Selbst- und Fremdeinschätzung von OPD-KJ-Konflikten durch Patient_innen und Therapeut_innen [Psychodynamic conflicts: Development of a questionnaire for the assessment by patients and their therapists]. *Z. Kinder- Jugendpsychiatrie Psychother.* **2021**, *49*, 361–376. [CrossRef]
26. Cropp, C.; Claaßen, B. Reliabilität und Validität des OPD-KJ-Konfliktfragebogens bei stationär behandelten Jugendlichen [Reliability and validity of the OPD-CA conflict questionnaire in an adolescent inpatient sample]. *Z. Kinder- Jugendpsychiatrie Psychother.* **2021**, *49*, 377–385. [CrossRef]
27. Schauenburg, H.; Willenborg, V.; Sammet, I.; Ehrenthal, J.C. Self-reported defence mechanisms as an outcome measure in psychotherapy: A study on the German version of the Defence Style Questionnaire DSQ 40. *Psychol. Psychother.* **2007**, *80*, 355–366. [CrossRef]
28. Sarrar, L.; Goth, K. Defense mechanisms reloaded in the light of impaired personality functioning: An attempt of clarification and simplification resulting in the DSQ-22-A for adolescents. *Front. Psychiatry* **2022**, *13*, 866837. [CrossRef]
29. Gräfe, K.; Zipfel, S.; Herzog, W.; Löwe, B. Screening psychischer Störungen mit dem "Gesundheitsfragebogen für Patienten (PHQ-D)" [Screening for mental disorders using the "Patient Health Questionnaire (PHQ-D)"]. *Diagnostica* **2004**, *50*, 171–181. [CrossRef]
30. Cohen, J. *Statistical Power Analysis for the Behavioral Sciences*; Lawrence Erlbaum: Hillsdale, NJ, USA, 1988.
31. Backhaus, K.; Erichson, B.; Plinke, W.; Weiber, R. *Multivariate Analysemethoden: Eine Anwendungsorientierte Einführung [Multivariate Analysis Methods: An Application-Oriented Introduction]*, 11th ed.; Springer: Berlin/Heidelberg, Germany, 2006.
32. Deng, J.; Zhou, F.; Hou, W.; Heybati, K.; Lohit, S.; Abbas, U.; Silver, Z.; Wong, C.Y.; Chang, O.; Huang, E.; et al. Prevalence of mental health symptoms in children and adolescents during the COVID-19 pandemic: A meta-analysis. *Ann. N. Y. Acad. Sci.* **2023**, *1520*, 53–73. [CrossRef]
33. Octavius, G.S.; Silviani, F.R.; Lesmandjaja, A.; Angelina; Juliansen, A. Impact of COVID-19 on adolescents' mental health: A systematic review. *Middle East. Curr. Psychiatry* **2020**, *27*, 72. [CrossRef]
34. Nearchou, F.; Flinn, C.; Niland, R.; Subramaniam, S.S.; Hennessy, E. Exploring the Impact of COVID-19 on Mental Health Outcomes in Children and Adolescents: A Systematic Review. *Int. J. Environ. Res. Public Health* **2020**, *17*, 8479. [CrossRef]
35. Evans, S.; Mikocka-Walus, A.; Klas, A.; Karantzas, G.; Westrupp, E.M. From "it has stopped our lives" to "spending more time together has strengthened bonds": The varied experiences of Australian families during COVID-19. *Front. Psychol.* **2020**, *11*, 588667. [CrossRef]
36. O'Reilly, J. Psychodynamic Theory: The development of a model of the mind. In *Seminars in the Psychotherapies*; Gibbons, R., O'Reilly, J., Eds.; Cambridge University Press: Cambridge, UK, 2021; pp. 3–20.
37. Khazand, S.; Riley, T.N.; Whitener, M.P.; Zapolski, T.C.B. Who am I: Considerations for adolescent development during a pandemic. *J. Ment. Health Soc. Behav.* **2021**, *3*, 143. [CrossRef]
38. Klotter, C. (Ed.) Essstörungen und Identität [Eating disorders and identity]. In *Identitätsbildung über Essen*; Springer: Berlin/Heidelberg, Germany, 2016; pp. 19–27.
39. Vismara, L.; Lucarelli, L.; Sechi, C. Attachment style and mental health during the later stages of COVID-19 pandemic: The mediation role of loneliness and COVID-19 anxiety. *BMC Psychol.* **2022**, *10*, 62. [CrossRef]
40. Coulombe, B.R.; Yates, T.M. Attachment security predicts adolescents' prosocial and health protective responses to the COVID-19 pandemic. *Child. Dev.* **2022**, *93*, 58–71. [CrossRef]
41. Dal Santo, T.; Sun, Y.; Wu, Y.; He, C.; Wang, Y.; Jiang, X.; Li, K.; Bonardi, O.; Krishnan, A.; Boruff, J.T.; et al. Systematic review of mental health symptom changes by sex or gender in early-COVID-19 compared to pre-pandemic. *Sci. Rep.* **2022**, *12*, 11417. [CrossRef]
42. Prowse, R.; Sherratt, F.; Abizaid, A.; Gabrys, R.L.; Hellemans, K.G.; Patterson, Z.R.; McQuaid, R.J. Coping with the COVID-19 pandemic: Examining gender differences in stress and mental health among university students. *Front. Psychiatry* **2021**, *12*, 650759. [CrossRef]
43. Duan, L.; Shao, X.; Wang, Y.; Huang, Y.; Miao, J.; Yang, X.; Zhu, G. An investigation of mental health status of children and adolescents in China during the outbreak of COVID-19. *J. Affect. Disord.* **2020**, *275*, 112–118. [CrossRef]
44. Zhou, S.-J.; Zhang, L.-G.; Wang, L.-L.; Guo, Z.-C.; Wang, J.-Q.; Chen, J.-C.; Liu, M.; Chen, X.; Chen, J.-X. Prevalence and socio-demographic correlates of psychological health problems in Chinese adolescents during the outbreak of COVID-19. *Eur. Child. Adolesc. Psychiatry* **2020**, *29*, 749–758. [CrossRef]

45. Kessler, R.C.; Petukhova, M.; Sampson, N.A.; Zaslavsky, A.M.; Wittchen, H.-U. Twelve-month and lifetime prevalence and lifetime morbid risk of anxiety and mood disorders in the United States. *Int. J. Methods Psychiatr. Res.* **2012**, *21*, 169–184. [CrossRef]
46. Demetriou, C.; Ozer, B.U.; Essau, C.A. Self-report questionnaires. In *The Encyclopedia of Clinical Psychology*; Cautin, R.L., Lilienfeld, S.O., Eds.; John Wiley & Sons: West Sussex, UK, 2015.

children

MDPI

Article

The Associations between Lifestyle Habits and Anxiety: A Prospective Study on Adolescents One Year after the Outbreak of the COVID-19 Pandemic

Laura Pedrini [1,*], Serena Meloni [1], Julia Dawson [1], Andrea Geviti [2], Natale Salvatore Bonfiglio [2], Anna Cattaneo [3,4] and Roberta Rossi [1]

1 Unit of Psychiatry, IRCCS Istituto Centro San Giovanni di Dio Fatebenefratelli, Via Pilastroni, 4, 25125 Brescia, Italy; smeloni@fatebenefratelli.eu (S.M.); jdawson@fatebenefratelli.eu (J.D.); rrossi@fatebenefratelli.eu (R.R.)
2 Statistics Service, IRCCS Istituto Centro San Giovanni di Dio Fatebenefratelli, Via Pilastroni, 4, 25125 Brescia, Italy; ageviti@fatebenefratelli.eu (A.G.); nbonfiglio@fatebenefratelli.eu (N.S.B.)
3 Biological Psychiatry Unit, IRCCS Istituto Centro San Giovanni di Dio Fatebenefratelli, Via Pilastroni, 4, 25125 Brescia, Italy; acattaneo@fatebenefratelli.eu
4 Department of Pharmacological and Biomolecular Sciences, University of Milan, 20133 Milan, Italy
* Correspondence: lpedrini@fatebenefratelli.eu

Abstract: Changes in lifestyle during the pandemic may have predisposed adolescents to vulnerability to poor mental health. This study aims to evaluate these changes and their association with the course of anxiety. A prospective study was conducted with 153 participants (16 years old, 72% female) who were assessed before the pandemic (T0, November 2019–January 2020) and one year later (T1, April–May 2021). Lifestyle habits (free-time activities, maladaptive behaviors, sleep, screen use) and anxiety were measured. Data concerning experiences related to COVID-19 and family relations during lockdown were collected. A worsening in lifestyle habits and anxiety was found. Of note, the pattern of associations between lifestyle habits and anxiety was quite different in the two time-points, suggesting that the purpose and the impact of some habits may be changed after the pandemic. Regression analyses showed that increases in anxiety were associated with increases in sleep problems, heightened efforts to reduce screen time, and loneliness. Pathway analysis revealed the absence of cross-lagged effects among anxiety, screen use, and sleep, while concurrent associations between variables were found in both the assessments. These results suggest possible long-term effects of the pandemic. Risk-factors associated with the course of anxiety were identified among lifestyle habits, thus contributing to identifying targets for interventions.

Keywords: adolescents; lifestyle; sleep; internet use; screen use

Citation: Pedrini, L.; Meloni, S.; Dawson, J.; Geviti, A.; Bonfiglio, N.S.; Cattaneo, A.; Rossi, R. The Associations between Lifestyle Habits and Anxiety: A Prospective Study on Adolescents One Year after the Outbreak of the COVID-19 Pandemic. *Children* **2024**, *11*, 282. https://doi.org/10.3390/children11030282

Academic Editor: Annio Posar

Received: 29 January 2024
Revised: 20 February 2024
Accepted: 22 February 2024
Published: 25 February 2024

1. Introduction

The restrictions that have been formulated to contain the COVID-19 pandemic, such as social distancing, school closures, and halting sports activities, directly influenced adolescents' daily routines. Several studies that were conducted during the first phase of the pandemic reported changes in the lifestyles of adolescents such as worsened sleep quality [1–5], increased amount of time spent online [6–8], and modifications in dietary habits [9].

Overall, these changes in habits may have predisposed adolescents to a higher vulnerability to poor mental health because they may have affected relevant developmental milestones [10]. For instance, it is well known that relationships with peers play a crucial role in the process associated with identity formation; therefore, if these relationships are disrupted, it may lead to loneliness and emotional problems [11,12]. Moreover, a large body of research shows that sleep efficiency promotes emotion regulation, alertness, and impulse control, whereas sleep-related problems have been found to be associated with a decrease in positive affect and with an increase in anxiety [13]. Finally, the digital world offers a spectrum of benefits and drawbacks: on one hand, it serves as a crucial avenue for adolescents to connect with peers and momentarily divert their attention from challenges. On the other hand, certain individuals might encounter difficulties in effectively self-regulating their engagement with digital technologies, potentially exacerbating pre-existing mental health problems [14–16].

Epidemiological studies conducted during the pandemic reveal a rising trend in clinically heightened anxiety symptoms over time among adolescents, aligning with concerns about COVID-19's effects on this demographic [6,17,18]. At the same time, it has been pointed out that not all young people showed a deterioration in their psychological health during the pandemic [19–21]. Therefore, long-term, accurate monitoring to identify risk and resilience factors associated with different mental health paths has gained importance. In this framework, it could be useful to evaluate the association between adolescents' lifestyle habits and mental health conditions. However, most of the studies on this topic are cross-sectional, with only a limited number offering evidence clarifying the mechanisms behind the associations with mental health.

Based on this knowledge, we conducted a prospective study with the following aims: (i) to compare lifestyle habits (i.e., free-time activities, maladaptive behaviors, sleep quality, screen use) reported by students just before the pandemic and one year later. We hypothesized a trend of change in lifestyle habits as shown by previous studies conducted during the first lockdown; (ii) to evaluate the association across time between anxiety, lifestyle habits, and the variables related to the experience of the lockdown. We hypothesized that some habits became more frequent among adolescents and may not be distinguishable from those of adolescents with different levels of anxiety. Moreover, we hypothesized that changes in some behaviors are associated with changes in anxiety, thus helping to explain the different reactions to stress induced by the pandemic; (iii) to deepen the investigation concerning the relations between the course of sleep problems over time, screen use, and anxiety. Specifically, we tested the same model with both cross-lagged effects and concurrent associations among these variables, thus contributing to clarifying the association in such relevant domains.

2. Materials and Methods

This is a prospective study conducted as part of a randomized clinical trial (RCT) on emotional dysregulation in adolescents (ClinicalTrials.gov ID: NCT04349709). The study was proposed to a high school in a city located in Northern Italy. Each school independently selected at least two classes in the 10th grade to participate. Following the protocol of the initial study, students with certified diagnoses of autism spectrum disorders were excluded.

For the purpose of the present study, we included a sample assessed at two timepoints: a pre-COVID baseline (T0) with data collection from November 2019 to January 2020 and a 1-year follow-up assessment (T1) with data collection during April and May 2021. Specifically, the sample included 153 students from eight classes in three high-schools.

Both the assessments were arranged during regular school day hours: at T0 participants completed a paper-based battery of questionnaires at school, while at T1 they completed an online version of the same battery. The researchers were available for students during both the sessions.

This study was approved by the local ethics committee (approval n.40/2019 on 11 July 2019 and n.13/2021 on 26 February 2021) and was carried out following the ethical standards presented in the 1964 Declaration of Helsinki. Written informed consent was obtained both at T0 and T1 from all subjects.

The dataset of the current study is available in the Zenodo repository (Zenodo DOI 10.5281/zenodo.10090891).

2.1. Assessment

The following variables were assessed both at baseline (T0) and one year later (T1):

Free-time activities. Participants were asked to describe their free-time activities by reporting the amount of daily time (1) spent with friends, (2) watching TV, (3) reading, (4) and being inactive and (5) the amount of weekly time spent playing sports on a four-level scale (1 = not at all, 2 = up to 1 h, 3 = from 1 to 2 h, 4 = more than 2 h).

Maladaptive behaviors. Participants were asked to report the frequency of the following maladaptive behaviors: (a) binge eating, (b) binge drinking behaviors, and (c) cannabis use in three separate questions. Subjects could respond based on the following frequency scale: 0 = never; 1 = only one time; 2 = once a week; 3 = two or three times a week; 4 = at least 4 times a week.

Sleep quality. Participants were asked to describe their sleep quality by answering the following questions [22]: (1) Well rested = "How many of the past seven days did you get enough sleep so that you felt rested when you woke up in the morning?" (response options: from 0 to 7 days); (2) Early awakening = "In the past seven days, how often have you awakened too early in the morning and couldn't get back to sleep?" (response options: from 0 to 7 days); (3) Difficulties falling asleep = "In the past seven days, how often have you had an extremely hard time falling asleep?" (response options: from 0 to 7 days); (4) Problems with daytime sleepiness = "In the past seven days, how much of a problem have you had with sleepiness during your daytime activities?" (response options: 1 = no problem at all; 2 = a little problem; 3 = more than a little problem; 4 = a big problem; 5 = a very big problem).

Screen use. Participants were asked to describe their use of social media or the internet during last month by answering the following four questions: (1) How many hours do you usually use social media/internet per day? (0 = up to one hour; 1 = from 1 to 2 h; 2 = from 3 to 5 h; 3 = more than 5 h). (2) How frequently do you find yourself staying up late to spend more time online/on the internet? (3) How often have you found yourself having little time to study because of the time spent online/on the internet? (4) How often have you found yourself reducing the amount of time you spend online/on the internet? Response options for questions 2, 3, 4 were as follows: 0 = never, 1 = only once; 2 = once a week; 3 = at least twice a week. In order to better describe problems related to screen use we calculated the "Problematic screen use" variable. To this purpose, we considered both the amount of daily screen time and related problems. Specifically, participants spending more than 2 h online daily [23] and scoring "3" on at least one of questions 2, 3, or 4 were classified as having "problematic screen use".

Anxiety. The Screen for Child Anxiety-Related Emotional Disorders (SCARED) [24] was used to measure anxiety. The SCARED is a 38-item self-report questionnaire measuring anxiety with the following five subscales: panic disorder, generalized anxiety disorder, separation anxiety disorder, school anxiety, and social anxiety. Each item is rated on three-point scale (1 = almost never, 2 = sometimes, 3 = often). The sum of all the items results in the total score. The Italian version of the SCARED showed good internal consistency (Cronbach's alpha ranged from 0.66 to 0.94) [25].

Moreover, at T1 we also registered the following data:

COVID-19-related experience. Participants were asked to report whether they contracted COVID-19 (yes; no); whether anyone of their family members contracted COVID-19 (yes; no); whether anyone of their family members had been hospitalized due to COVID-19 (yes; no); and whether anyone of their family members died from COVID-19 (yes; no). Moreover, they rated how worried they were about contracting COVID-19 on a 0–10 visual analog scale.

Lockdown-related experience. Participants were asked to report whether during lockdown their parents stayed at home (yes; no); whether conflicts in the family increased (yes; no); and whether they felt alone (yes; no).

2.2. Data Analysis

Data was cleaned, coded, and evaluated using IBM SPSS Statistics, version 21 [26]. The Jamovi package was used only for the path analysis [27]. The reported confidence intervals were set at 95%. The assumptions of normality and multivariate normality were assessed by visual inspection of the QQ plots and density histograms for each variable. Durbin–Watson tests were used to detect autocorrelation of residuals, variation inflation factors (VIFs) and tolerance indexes were used for collinearity, and Levene's test was used for homoscedasticity. No significant violations of the assumption were found. The absolute and relative frequencies and the means and standard deviations were calculated for each of the sociodemographic and response variables. The Chi-square test was carried out for categorical variables in order to investigate differences between subgroups at baseline.

In order to achieve a global index of social media/internet use, a new variable called "screen use composite score" was created. It corresponds to the factor score obtained through the principal component analysis (PCA) considering the responses to questions n1, n2, and n3 on the screen use. Only one dimension was extracted (60.2% of explained variance for T0 and 64.3% for T1) with a Cronbach's alpha of 0.67 for T0 and 0.72 for T1.

In order to evaluate the difference between T0 and T1 for each variable of interest, a series of paired Student *t*-tests, for continuous variables, and McNemar tests, for categorical variables, were calculated.

In order to evaluate the change over time, a series of delta variables was created for each of the variables of interest (i.e., delta = variable at T1—variable at T0). Pearson's r was used to find correlation between the delta variables. A series of multivariate regression models were then created for those variables that were significantly correlated with the delta of anxiety's total score.

Finally, a path analysis was performed taking into account the third aim of the study, which was to determine the association between sleep, social media/internet use, and anxiety. Specifically, the following variables assessed at both T0 and T1 were used: (a) total anxiety score, (b) difficulty falling asleep, and (c) online composite score. For model performance, the following fit indexes were used: the Chi-square, the root mean square error approximation (RMSE), the standardized root mean square residual (SRMR), the comparative fit index (CFI), the Tucker–Lewis index (TLI), and the goodness of fit (GFI) [28,29].

3. Results

Participants were mainly female (72%), and the average age was 16 years old (16.1 ± 0.49). A total of 70 students (46%) reported that at least one of their family members contracted COVID-19. A total of 24 (16%) students reported that at least one of their family members had been hospitalized for COVID-19 treatment, and n = 15 (10%) students reported that at least one of their family members died of COVID-19. A total of 19 (12%) students contracted COVID-19. On average, participants expressed moderate levels of concern about the possibility of contracting the virus on a 0–10 visual analog scale (5.5 ± 2.1).

3.1. Lifestyle Habits and Anxiety across Time

As reported in Table 1, there was a decrease in the amount of time spent with friends ($p < 0.001$) and in reading ($p < 0.001$). The frequency of binge-eating episodes was higher at follow-up compared to the pre-pandemic period ($p = 0.003$). Moreover, there was an overall increase in difficulties falling asleep ($p = 0.036$) and daytime sleepiness ($p = 0.011$). The time spent online increased over time ($p < 0.001$). In particular, those that declared spending more than 2 h a day online were n = 94 (61%) and n = 122 (80%), respectively, at baseline and at follow-up (without considering online learning). After the pandemic, a relevant number of students (n = 43, 28.1%) stated that they spent more than 5 h online a day (without considering online learning). There was an increase in the proportion of adolescents frequently staying up late to spend more time online ($p < 0.001$), as well as trying to reduce the amount of time spent online ($p < 0.001$). Finally, the proportion of students classified as having "Problematic screen use" significantly increased over time (n = 59; 38.6% at T0; n = 93, 60.8% at T1, $p = < 0.001$).

Table 1. Lifestyle habits before the pandemic (T0) and one year later (T1).

	T0 N = 153	T1 N = 153	Test (DF)	*p*-Value [1]	Effect Size $
		Free-time activities			
Time spent with friends (1–4)	2.93 ± 0.96	2.42 ± 1.01	t = 9.64 (150)	$p < 0.001$	0.40
Time spent watching TV (1–4)	2.03 ± 0.90	1.94 ± 0.97	t = 1.25 (151)	$p = 0.214$	0.10
Time spent reading (1–4)	1.88 ± 0.87	1.69 ± 0.88	t = 3.1 (151)	$p = 0.002$	0.25
Time spent inactive (1–4)	2.31 ± 1.01	2.21 ± 1.05	t = 1.21 (150)	$p = 0.227$	0.10
Time spent playing sports (0–4)	2.16 ± 1.63	1.96 ± 1.55	t = 1.41 (152)	$p = 0.161$	0.11
		Maladaptive behaviors			
Binge drinking (0–4)	0.18 ± 0.47	0.16 ± 0.49	t = 0.52 (152)	$p = 0.607$	0.04
Binge eating (0–4)	0.12 ± 0.43	0.37 ± 0.97	t = −3 (152)	$p = 0.003$	0.24
Cannabis use (0–4)	0.11 ± 0.47	0.22 ± 0.67	t = −1.76 (152)	$p = 0.081$	0.14
		Sleep quality			
Enough sleep (0–7)	2.03 ± 1.9	1.77 ± 1.6	t = 1.61 (148)	$p = 0.110$	0.13
Early awakening (0–7)	1.14 ± 1.9	1.03 ± 1.6	t = 0.64 (152)	$p = 0.524$	0.05
Difficulties in falling asleep (0–7)	1.86 ± 2.2	2.30 ± 2.4	t = −2.11 (158)	$p = 0.036$	0.17
Daytime sleepiness (0–4)	1.05 ± 0.93	1.27 ± 1.02	t = −2.58 (152)	$p = 0.011$	0.21
		Screen use			
Time spent online (1–4)	2.63 ± 0.71	3.05 ± 0.75	t = −6.34 (152)	$p < 0.001$	0.51
Staying up too late to spend more time online (0–5)	1.37 ± 1.42	1.93 ± 1.35	t = −4.44 (152)	$p < 0.001$	1.21
Little time to study because of time spent online (0–5)	1.30 ± 1.32	1.71 ± 1.35	t = −3.39 (152)	$p < 0.001$	0.27
Attempts to reduce the amount of time online (0–4)	1.62 ± 1.35	2.21 ± 1.37	t = −4.86 (152)	$p < 0.001$	0.39
Problematic screen use	N = 59 (38.5%)	N = 93 (60.7%)	$X^2 = 23.13$ (1)	$p < 0.001$	0.37

[1] Paired *t*-test for ordinal variables, McNemar test for categorical variables. $: Effect size = Cohen's d for all variables except "Problematic screen use", for which Cramer's V was calculated.

Anxiety levels increased across time, as shown by the difference in the total SCARED mean score (64.9 ± 12.1 at T0 and 67 ± 12.6 at T1; $p = 0.002$) and in the following subscales: generalized anxiety (19.4 ± 3.9 at T0 and 20.4 ± 4.2 at T1; $p < 0.001$) and school anxiety (6.8 ± 1.9 at T0 and 7.6 ± 2.3 at T1; $p < 0.001$). In contrast, we did not find relevant differences for panic symptoms (20.2 ± 5.8 at T0 and 20.4 ± 5.6 at T1; $p = 0.383$), separation anxiety (11.2 ± 2.5 at T0 and 11.1 ± 2.6 at T1; $p = 0.574$), and social anxiety (7.6 ± 2.4 at T0 and 7.7 ± 2.6 at T1; $p = 0.400$).

3.2. Associations between Anxiety and Current Lifestyle Habits

Table 2 shows the associations between anxiety and lifestyle habits that were measured at the same time. Findings show that, before the COVID-19 pandemic, anxiety was associated with time spent with friends, sports practice, binge-eating episodes, and sleep-related problems. These associations were also confirmed at the 1-year follow up. Moreover, at T0, higher levels of anxiety were associated with the number of hours spent inactive and the number of hours spent online, whereas our findings at the 1-year follow up no longer indicated an association between these behaviors and levels of anxiety. An inverse correlation between anxiety and the feeling of being well-rested upon waking up emerged after the pandemic. Finally, after the pandemic higher anxiety levels were significantly associated with dysfunctional internet use (i.e., staying up late to spend more time online, reduced time to study, ineffective attempts to reduce screen use), rather than with the number of hours spent online as found at baseline.

Table 2. Correlations between anxiety and current lifestyle habits (N = 153).

	Anxiety at T0		Anxiety at T1	
	r	*p* Value	r	*p* Value
Time spent with friends	−0.164	$p = 0.043$	−0.213	$p = 0.008$
Time spent watching TV	0.105	$p = 0.154$	0.044	$p = 0.584$
Time spent reading	0.168	$p = 0.035$	0.073	$p = 0.369$
Time spent inactive	0.235	$p = 0.003$	−0.037	$p = 0.650$
Time spent playing sports	−0.296	$p = 0.001$	−0.282	$p < 0.001$
Binge drinking	−0.076	$p = 0.370$	−0.150	$p = 0.065$
Binge eating	0.174	$p = 0.046$	0.362	$p < 0.001$
Cannabis use	0.064	$p = 0.474$	0.008	$p = 0.926$
Enough sleep	−0.072	$p = 0.570$	−0.283	$p < 0.001$
Early awakening	0.239	$p = 0.003$	0.191	$p = 0.018$
Difficulties in falling asleep	0.215	$p = 0.015$	0.332	$p < 0.001$
Daytime sleepiness	0.247	$p = 0.002$	0.383	$p < 0.001$
Time spent online	0.196	$p = 0.013$	0.070	$p = 0.391$
Staying up late to spend more time online	0.145	$p = 0.061$	0.181	$p = 0.025$
Little time to study because of time spent online	0.097	$p = 0.197$	0.196	$p = 0.015$
Attempts to reduce the amount of time spent online	0.217	$p = 0.011$	0.218	$p = 0.007$

3.3. *Factors Associated with the Change in Anxiety Levels over Time*

In order to test possible predictors of longitudinal change in anxiety symptoms, separated models of regression analysis were prepared. We considered as potential predictors both the changes in lifestyle habits (delta variables) and the factors related to the experience of the first lockdown. Specifically, among lifestyle habits we considered only the variables significantly correlated with changes in anxiety, which were the variations in difficulty in falling asleep ($r = 0.32$; $p < 0.001$) and in attempts to reduce screen time use ($r = 0.24$; $p = 0.002$).

Our findings show that the increase in difficulties in falling asleep is associated with the increase in anxiety levels over time ($p < 0.001$). Moreover, participants reporting an increase in ineffective attempts to reduce the amount of time spent online were more likely to record an increase in anxiety ($p = 0.003$). Concerning the variables related to the experience of the first lockdown, we found no evidence for an effect of home climate (i.e., parents at home, increased family conflicts) on levels of anxiety. In contrast, having experienced loneliness during the first lockdown ($p = 0.027$) and having experienced the death of a family member ($p = 0.012$) were associated with increased anxiety levels over time. Subsequently, the statistically significant variables were simultaneously entered in the same model (Table 3). All of them, except for the one related to having experienced death of a family member due to COVID-19, continued to be significantly associated with the course of anxiety symptoms. Specifically, it was found that the increase in difficulties in falling asleep ($p < 0.001$; standardized $\beta = +0.27$) and in the attempts to reduce the amount of time spent online ($p = 0.005$; standardized $\beta = +0.21$) and having experienced feelings of loneliness during lockdown ($p = 0.011$; standardized $\beta = -0.19$) were predictors of worsened anxiety symptoms over time, even if the standardized coefficients suggest small effects.

Table 3. Variables associated with the change in anxiety level across time; multiple regression model (N = 153) *.

	p	Beta	Standardized β	Adjusted R2
Δ Difficulties in falling asleep	<0.001	+0.86	+0.27	
Δ Attempts to reduce screen time	0.005	+1.13	+0.21	0.199
Loneliness	0.011	−5.05	−0.19	
COVID-related experiences in the family	0.05			

* a positive value of Δ SCARED denotes an improvement in the SCARED total score across time, while a negative value of Δ SCARED denotes a worsening in the SCARED total score across time.

3.4. Cross-Lagged and Concurrent Associations among Sleep, Screen Use, and Anxiety

Based on the significant associations found among sleep problems, screen use, and anxiety, a pathway analysis was conducted in order to explore their relationships at the two time-points. Specifically, we tested both concurrent association and cross-lagged effects. As shown in Figure 1, the results revealed no significant cross-lagged associations, meaning that none of the variables measured at T0 was associated with the others measured at T1 (for example: anxiety at T0 was not associated with screen use, nor with sleep at T1). Concurrent associations were found between anxiety and sleep in both the assessments; that is, anxiety and sleep were associated at both time points of the assessment. while screen use was associated with anxiety only before the pandemic. According to recognized standards, the model appears to be very acceptable based on the fit index.

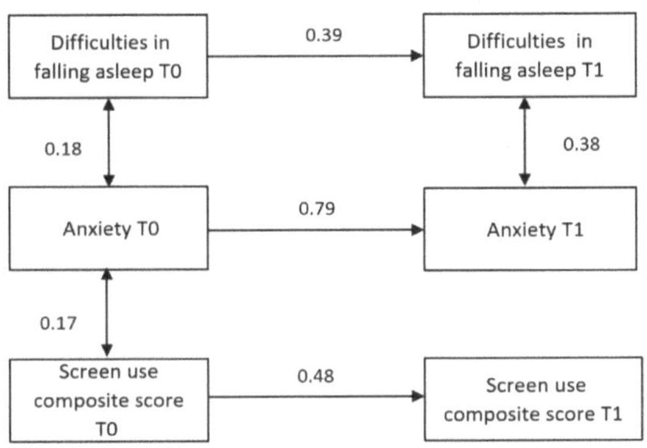

Figure 1. Path model with beta coefficients. Beta values are shown for each unidirectional and bidirectional interaction in arrows. Normality and Mardia tests revealed only skewness and not kurtosis tendencies (Skewness: 2 = 85; df = 56; $p = 0.007$; Kurtosis: z = 0.81; $p = 0.419$); hence, the robust maximum likelihood technique was employed to estimate SEM parameters. Several fit indexes revealed the goodness of the model: the Chi-square value was 11.5 (df = 9; $p = 0.244$); the RMSEA was 0.042; the standardized version (SRMSE) was 0.055; the Tucker–Lewis Index (TLI) was 0.990; the comparative fit index (CFI) was 0.983; and the GFI was 0.998.

4. Discussion

The present data refers to a period with less stringent restrictions, as it was one year after the outbreak of COVID-19; therefore they seem to suggest a possible long-term effect of the pandemic on adolescents' daily routines. Indeed, similarly to the first wave of the pandemic [1–8], we found a decrease in time spent with friends and reading, whereas sleep-related problems, binge-eating episodes, and screen use increased over time. Of note, staying up too late to spend more time online was the behavior that showed the strongest effect size, thus suggesting that sleep quality may have been particularly affected during this period.

Interestingly, the variables associated with anxiety were slightly different in the two time-points, suggesting that certain behaviors may have taken on different purposes and consequences in daily routines. Before the pandemic, anxiety showed an inverse correlation with time spent with friends and engagement in sports activities. Conversely, it was directly associated with time spent inactive, binge-eating episodes, sleep-related problems, and online time. These findings align with the well-established literature on the relationship between lifestyle and mental health [11–14] and emphasize the importance of interventions aimed at promoting functional habits to support mental health. After the pandemic, some behaviors (i.e., time spent inactive and the amount of time spent online) were no longer significantly associated with anxiety. It is important to note that during the lockdown period, increases in inactivity and heightened screen use became common behaviors among adolescents. As a result, it seems that these behaviors no longer serve as distinguishing factors for the emotional state of youths, as they did prior to the pandemic. In particular, social media may have represented a vital source for fostering peer connections and providing valuable support, especially in the midst of lockdown restrictions. Of note, a study conducted during the pandemic showed that screen time dedicated to fostering connections was associated with reduced feelings of loneliness and heightened levels of overall well-being [9].

Gaining insight into the mechanisms that underscore the connection between lifestyle habits and anxiety holds significance, as it facilitates the identification of potential intervention targets [30]. In our study, the increase in anxiety over time was predicted by having experienced feelings of loneliness during lockdown, loss of a family member due to COVID-19, difficulties in falling asleep, and difficulties in curtailing screen use. These results lead us to hypothesize that these variables may have served as pandemic-induced risk factors, potentially contributing to emotional challenges. The school closure and the resultant inactivity due to the restrictions have undeniably led to alterations in sleep patterns, which may have in turn negatively impacted levels of anxiety. Conversely, there is a plausible scenario wherein the overarching uncertainty stemming from the pandemic could induce a pervasive sense of generalized anxiety or heightened alertness, consequently leading to disruptions in sleep patterns. It is well recognized that sleep-related symptoms and anxiety share an intrinsic connection, and often they overlap [31]. To our knowledge, the only study aiming to investigate the direction of the relationship between sleep and anxiety during the pandemic found support of a bidirectional association between these variables [1]. Interestingly, the results of our pathway analysis showed similar evidence. Based on these findings, interventions designed to cultivate awareness regarding sleep patterns and enhance strategies for upholding a functional daily routine could be useful. In reinforcement of this proposal, we observed that feeling well-rested upon morning awakening exhibited a significant association with reduced anxiety levels exclusively in the second assessment. This highlights the noteworthy influence of good sleep during challenging circumstances such as the pandemic, as it potentially mitigates the adverse impact of stressful events. Indeed, there is evidence documenting the influence of the quality of sleep on effective emotion regulation and positive affect [13].

Another main finding of the present study refers to problematic screen use that emerges as a factor associated with anxiety; this is interesting since the influence of screen use on mental health remains a topic of contention. Increased use of screen technology has been found to be associated with heightened mental health problems [32]. Conversely, a recent review encompassing longitudinal studies reached the conclusion that there exists only a minor direct impact of screen use on psychopathological symptoms [33]. Indeed, according to the "displacement hypothesis", adverse consequences arise when screen use substitutes other healthy behaviors, such as social interaction, functional problem-solving, and functional lifestyle habits [8,14,16]. Consistent with this theory, after the pandemic our sample yielded no correlation between anxiety and the number of hours spent online, whereas we observed an association between anxiety and the presence of dysfunctional screen use (i.e., staying up late to spend more time online, time to study negatively affected by the amount of time spent online, ineffective attempts to reduce amount of time spent online). We can speculate that challenges in effectively regulating screen use, rather than the mere duration of time spent online, might play a role in exacerbating anxiety. In line with this interpretation, a review of the studies conducted on clinical samples of youth patients has demonstrated that certain young individuals might encounter challenges in self-regulating their screen-time, which in turn could amplify psychological distress [15]. Similarly, another study showed that the impact of screen use on mental health may be mediated by coping strategies. Adolescents who tend to adopt functional coping strategies appear to exhibit greater resilience against the depressive symptoms associated with screen time [16]. Considering this evidence, the findings of the present study are valuable in identifying a risk factor for poor mental health in adolescents during the pandemic. The study highlights a potential unfavorable condition for a significant portion of the sample, as we observed an increase from 39% to 61% of participants with problematic screen use.

Lastly, loneliness emerges as another pivotal factor for higher levels of anxiety, and this finding is in line with several studies conducted on young people during the COVID-19 pandemic [34]. According to the cognitive model of anxiety [35], perceiving a lack of support intensifies feelings of threat amid stress-inducing circumstances, and this may explain the association between feelings of loneliness and anxiety. Loneliness is an unpleasant and distressing condition that results from deficiencies in an individual's social connections, but it is also a subjective experience that refers to cognitive processes [36]. In this regard, in a large sample of adolescents it was found that the association between loneliness and psychopathological symptoms during lockdown was independent from social contacts, whereas it was significantly associated with rejection sensitivity (i.e., the inclination to anticipate social rejection with anxiety, promptly perceive it, and react excessively) [11]. The results of our study contribute to the existing literature by showing the importance of cognitive processes; indeed, we found that higher levels of anxiety were associated with feelings of loneliness during lockdown, whereas the variables closely tied to social interactions at home (i.e., having parents at home and family conflicts at home during lockdown) did not exhibit any significant association with anxiety. Of course, we cannot exclude that the prohibition of meeting friends, rather than the quality of family relationships, may have contributed to the experience of loneliness. However, it is also possible that our data underscores the importance of intrapersonal variables in mediating the impact of loneliness, leading to hypotheses of intervention. Specifically, since negative interpretation bias may serve as a mechanism through which loneliness becomes a precursor to emotional difficulties, interventions aimed at challenging these negative interpretations could be useful in mitigating mental health symptoms [37].

Strengths and Limitations

The present study has several strengths, with its prospective design and one-year follow-up being foremost among them. Additionally, the response rate enhances the reliability of the findings. The sample was sourced from a homogeneous geographical area which experienced a significant impact from COVID-19. The assessment of internet use in our study provides more comprehensive information compared to existing studies conducted during pandemic. While many previous studies only considered the number of hours spent online, our assessment also evaluated the impact on personal functioning. Finally, while the analyses offer valuable insights into the potential influence of the pandemic on the course of adolescents' anxiety, it is crucial to interpret these findings within the framework of the study's limitations. Indeed, it is important to acknowledge that the convenience-based sampling approach we employed, along with the high proportion of female participants, may limit the generalizability of our findings to the broader population of adolescents. Moreover, the lack of a sample size calculation to assess the power of the analysis and the small effects observed for some of the factors predicting the course of anxiety should be taken into consideration. Additionally, while loneliness emerged as a significant factor influencing mental health, it is worth noting that it was assessed using a single item. Lastly, it is important to emphasize that further studies are necessary to comprehensively capture the impact of the pandemic on adolescents' wellbeing, including aspects such as coping styles and emotional regulation strategies.

5. Conclusions

This study is among the limited number of prospective investigations featuring a 1-year follow-up. It sheds light on the increase in anxiety levels and the enduring prevalence of negative lifestyle habits, even during a period characterized by less stringent restrictions. The lifestyle habits associated with anxiety were slightly different at the two assessment points; therefore, future studies should consider that certain behaviors may have taken on different purposes and consequences in daily life. Specifically, after the pandemic, some behaviors (i.e., time spent inactive and the amount of time spent online) were no longer associated with anxiety, while quality of sleep became more strictly associated with anxiety. Moreover, at follow-up, dysfunctional internet use, rather than the sheer number of hours spent online, exhibited an association with anxiety levels. Potential risk factors that might have influenced the course of anxiety were identified, specifically, the increase in problematic screen use, sleep-related problems, and feelings of loneliness during lockdown. Overall, these data support the importance of interventions targeting these domains to improve adolescents' mental health.

Author Contributions: L.P., A.C. and R.R. conceived the study; L.P., S.M. and J.D. collected data; S.M. performed data entry and the data quality check; A.G. and N.S.B. performed statistical analyses; L.P. drafted the first version of the manuscript; and J.D. performed English language editing. All authors have read and agreed to the published version of the manuscript.

Funding: This study was supported by the Italian Ministry of Health (GR-2018-12366754, Ricerca Corrente).

Institutional Review Board Statement: The study was conducted in accordance with the Declaration of Helsinki and approved by the local ethics committee (approval n.40/2019 on 11 July 2019 and n.13/2021 on 26 February 2021).

Informed Consent Statement: Written informed consent was obtained from all subjects involved in the study.

Data Availability Statement: The data presented in this study are openly available in Zenodo repository at DOI 10.5281/zenodo.10090891.

Acknowledgments: We extend our heartfelt gratitude to the following schools situated in Brescia (Italy): Istituto Maddalena di Canossa, Liceo delle Scienze Umane Fabrizio De André, and Liceo Statale Leonardo for their invaluable collaboration. Our sincere appreciation also goes to the students who willingly took part in the study and diligently filled out the questionnaires. Additionally, we extend our thanks to the teachers who collaborated with the study.

Conflicts of Interest: The authors declare no conflicts of interest.

References

1. Wang, W.; Guo, Y.; Du, X.; Li, W.; Wu, R.; Guo, L.; Lu, C. Associations Between Poor Sleep Quality, Anxiety Symptoms, and Depressive Symptoms Among Chinese Adolescents Before and During COVID-19: A Longitudinal Study. *Front. Psychiatry* **2022**, *12*, 786640. [CrossRef] [PubMed]
2. Bacaro, V.; Meneo, D.; Curati, S.; Buonanno, C.; De Bartolo, P.; Riemann, D.; Mancini, F.; Martoni, M.; Baglioni, C. The impact of COVID-19 on Italian adolescents' sleep and its association with psychological factors. *J. Sleep Res.* **2022**, *31*, e13689. [CrossRef] [PubMed]
3. Dragun, R.; Veček, N.; Marendić, M.; Pribisalić, A.; Đivić, G.; Cena, H.; Polašek, O.; Kolčić, I. Have Lifestyle Habits and Psychological Well-Being Changed among Adolescents and Medical Students Due to COVID-19 Lockdown in Croatia? *Nutrients* **2020**, *13*, 97. [CrossRef] [PubMed]
4. Zhao, J.; Xu, J.; He, Y.; Xiang, M. Children and adolescents' sleep patterns and their associations with mental health during the COVID-19 pandemic in Shanghai, China. *J. Affect. Disord.* **2022**, *301*, 337–344. [CrossRef] [PubMed]
5. Illingworth, G.; Mansfield, K.L.; Espie, C.A.; Fazel, M.; Waite, F. Sleep in the time of COVID-19: Findings from 17000 school-aged children and adolescents in the UK during the first national lockdown. *Sleep Adv.* **2022**, *3*, zpab021. [CrossRef] [PubMed]
6. Pustake, M.; Mane, S.; Ganiyani, M.A.; Mukherjee, S.; Sayed, M.; Mithbavkar, V.; Memon, Z.; Momin, A.S.; Deshmukh, K.; Chordia, A.; et al. Have the COVID-19 pandemic and lockdown affected children's mental health in the long term? A repeated cross-sectional study. *BMJ Open* **2022**, *12*, e058609. [CrossRef]
7. Ozturk, F.O.; Ayaz-Alkaya, S. Internet addiction and psychosocial problems among adolescents during the COVID-19 pandemic: A cross-sectional study. *Arch. Psychiatr. Nurs.* **2021**, *35*, 595–601. [CrossRef] [PubMed]
8. Chen, Q.; Dai, W.; Li, G.; Ma, N. The impact of screen time changes on anxiety during the COVID-19 pandemic: Sleep and physical activity as mediators. *Sleep Biol. Rhythm.* **2022**, *20*, 521–531. [CrossRef]
9. Li, S.H.; Beames, J.R.; Newby, J.M.; Maston, K.; Christensen, H.; Werner-Seidler, A. The impact of COVID-19 on the lives and mental health of Australian adolescents. *Eur. Child Adolesc. Psychiatry* **2022**, *31*, 1465–1477. [CrossRef]
10. Garagiola, E.R.; Lam, Q.; Wachsmuth, L.S.; Tan, T.Y.; Ghali, S.; Asafo, S.; Swarna, M. Adolescent Resilience during the COVID-19 Pandemic: A Review of the Impact of the Pandemic on Developmental Milestones. *Behav. Sci.* **2022**, *12*, 220. [CrossRef]

11. Minihan, S.; Orben, A.; Songco, A.; Fox, E.; Ladouceur, C.D.; Mewton, L.; Moulds, M.; Pfeifer, J.H.; Van Harmelen, A.-L.; Schweizer, S. Social determinants of mental health during a year of the COVID-19 pandemic. *Dev. Psychopathol.* **2022**, *35*, 1701–1713. [CrossRef] [PubMed]
12. Orben, A.; Tomova, L.; Blakemore, S.-J. The effects of social deprivation on adolescent development and mental health. *Lancet Child Adolesc. Health* **2020**, *4*, 634–640. [CrossRef] [PubMed]
13. Guldner, S.; Sarvasmaa, A.S.; Lemaître, H.; Massicotte, J.; Vulser, H.; Miranda, R.; Frère, P.B.; Filippi, I.; Penttilä, J.; Banaschewski, T.; et al. Longitudinal associations between adolescent catch-up sleep, white-matter maturation and internalizing problems. *Dev. Cogn. Neurosci.* **2023**, *59*, 101193. [CrossRef] [PubMed]
14. Alonzo, R.; Hussain, J.; Stranges, S.; Anderson, K.K. Interplay between social media use, sleep quality, and mental health in youth: A systematic review. *Sleep Med. Rev.* **2021**, *56*, 101414. [CrossRef] [PubMed]
15. Kostyrka-Allchorne, K.; Stoilova, M.; Bourgaize, J.; Rahali, M.; Livingstone, S.; Sonuga-Barke, E. Review: Digital experiences and their impact on the lives of adolescents with pre-existing anxiety, depression, eating and nonsuicidal self-injury conditions—A systematic review. *Child Adolesc. Ment. Health* **2023**, *28*, 22–32. [CrossRef] [PubMed]
16. Hökby, S.; Westerlund, J.; Alvarsson, J.; Carli, V.; Hadlaczky, G. Longitudinal Effects of Screen Time on Depressive Symptoms among Swedish Adolescents: The Moderating and Mediating Role of Coping Engagement Behavior. *Int. J. Environ. Res. Public Health* **2023**, *20*, 3771. [CrossRef] [PubMed]
17. Newlove-Delgado, T.; Russell, A.E.; Mathews, F.; Cross, L.; Bryant, E.; Gudka, R.; Ukoumunne, O.C.; Ford, T.J. Annual Research Review: The impact of COVID-19 on psychopathology in children and young people worldwide: Systematic review of studies with pre- and within-pandemic data. *J. Child Psychol. Psychiatry* **2023**, *64*, 611–640. [CrossRef]
18. Racine, N.; McArthur, B.A.; Cooke, J.E.; Eirich, R.; Zhu, J.; Madigan, S. Global Prevalence of Depressive and Anxiety Symptoms in Children and Adolescents During COVID-19: A Meta-analysis. *JAMA Pediatr.* **2021**, *175*, 1142–1150. [CrossRef]
19. Grey, I.; Arora, T.; Thomas, J.; Saneh, A.; Tohme, P.; Abi-Habib, R. The role of perceived social support on depression and sleep during the COVID-19 pandemic. *Psychiatry Res.* **2020**, *293*, 113452. [CrossRef]
20. Pedrini, L.; Meloni, S.; Lanfredi, M.; Ferrari, C.; Geviti, A.; Cattaneo, A.; Rossi, R. Adolescents' mental health and maladaptive behaviors before the COVID-19 pandemic and 1-year after: Analysis of trajectories over time and associated factors. *Child Adolesc Psychiatry Ment. Health* **2022**, *16*, 42. [CrossRef]
21. She, R.; Wong, K.; Lin, J.; Zhang, Y.; Leung, K.; Yang, X. Profiles of stress and coping associated with mental, behavioral, and internet use problems among adolescents during the COVID-19 pandemic: A stratified random sampling and cluster analysis. *Front. Public Health* **2022**, *10*, 826911. [CrossRef] [PubMed]
22. Alamir, Y.A.; Zullig, K.J.; Wen, S.; Montgomery-Downs, H.; Kristjansson, A.L.; Misra, R.; Zhang, J. Association Between Nonmedical Use of Prescription Drugs and Sleep Quality in a Large College Student Sample. *Behav. Sleep Med.* **2019**, *17*, 470–480. [CrossRef] [PubMed]
23. Gentile, D.A.; Bailey, K.; Bavelier, D.; Brockmyer, J.F.; Cash, H.; Coyne, S.M.; Doan, A.; Grant, D.S.; Green, C.S.; Griffiths, M.; et al. Internet Gaming Disorder in Children and Adolescents. *Pediatrics* **2017**, *140* (Suppl. S2), S81–S85. [CrossRef]
24. Birmaher, B.; Brent, D.A.; Chiappetta, L.; Bridge, J.; Monga, S.; Baugher, M. Psychometric properties of the Screen for Child Anxiety Related Emotional Disorders (SCARED): A replication study. *J. Am. Acad. Child Adolesc. Psychiatry* **1999**, *38*, 1230–1236. [CrossRef] [PubMed]
25. Crocetti, E.; Hale, W.W., 3rd; Fermani, A.; Raaijmakers, Q.; Meeus, W. Psychometric properties of the Screen for Child Anxiety Related Emotional Disorders (SCARED) in the general Italian adolescent population: A validation and a comparison between Italy and The Netherlands. *J. Anxiety Disord.* **2009**, *23*, 824–829. [CrossRef]
26. IBM Corp. *IBM SPSS Statistics for Windows*; Version 21.0; IBM Corp.: Armonk, NY, USA, 2012.
27. The Jamovi Project. Jamovi [Computer Software], Version 2.3. 2022. Available online: https://www.jamovi.org (accessed on 15 March 2023).
28. Kline, R.B. Assumptions in structural equation modeling. In *Handbook of Structural Equation Modeling*; Guilford Press: New York, NY, USA, 2012; pp. 111–125.
29. Schreiber, J.B.; Nora, A.; Stage, F.K.; Barlow, E.A.; King, J. Reporting structural equation modeling and confirmatory factor analysis results: A review. *J. Educ. Res.* **2006**, *99*, 323–338. [CrossRef]
30. Kara, A. A review of childhood anxiety. *J. Clin. Trials Exp. Investig.* **2022**, *1*, 64–70.
31. American Psychiatric Association. Sleep-Wake Disorders. In *Diagnostic and Statistical Manual of Mental Disorders*, 5th ed.; text rev.; American Psychiatric Association: Washington, DC, USA, 2022.
32. Boers, E.; Afzali, M.H.; Newton, N.; Conrod, P. Association of Screen Time and Depression in Adolescence. *JAMA Pediatr.* **2019**, *173*, 853–859. [CrossRef] [PubMed]
33. Tang, S.; Werner-Seidler, A.; Torok, M.; Mackinnon, A.J.; Christensen, H. The relationship between screen time and mental health in young people: A systematic review of longitudinal studies. *Clin. Psychol. Rev.* **2021**, *86*, 102021. [CrossRef] [PubMed]
34. Loades, M.E.; Chatburn, E.; Higson-Sweeney, N.; Reynolds, S.; Shafran, R.; Brigden, A.; Linney, C.; McManus, M.N.; Borwick, C.; Crawley, E. Rapid Systematic Review: The Impact of Social Isolation and Loneliness on the Mental Health of Children and Adolescents in the Context of COVID-19. *J. Am. Acad. Child Adolesc. Psychiatry* **2020**, *59*, 1218–1239.e3. [CrossRef] [PubMed]
35. Clark, D.A.; Beck, A.T. *Cognitive Therapy of Anxiety Disorders: Science and Practice*; Guilford Press: New York, NY, USA, 2010.

36. Perlman, D.; Peplau, L.A. *Loneliness: A Sourcebook of Current Theory, Research, and Therapy*; Wiley Interscience: New York, NY, USA, 1982.
37. Gardner, A.A.; Zimmer-Gembeck, M.J.; Modecki, K. A longitudinal model of rejection sensitivity and internalizing symptoms: Testing emotion regulation deficits as a mechanism and outcome of symptoms. *J. Pers.* **2020**, *88*, 1045–1057. [CrossRef] [PubMed]

children

MDPI

Article

The Role of Affects and Emotional Styles in the Relationship Between Parents and Preschool Children †

Carolina Facci [1,*], Andrea Baroncelli [2] and Enrica Ciucci [1]

1 Department of Education, Languages, Interculture, Literatures and Psychology, University of Florence, Via di San Salvi 12, Complesso di San Salvi Padiglione 26, 50135 Florence, Italy; enrica.ciucci@unifi.it
2 Department of Philosophy, Social Sciences and Education, University of Perugia, Piazza Giuseppe Ermini 1, 06123 Perugia, Italy; andrea.baroncelli@unipg.it
* Correspondence: carolina.facci@unifi.it
† This is part of Dr. Facci's PhD thesis in Education Sciences and Psychology at the University of Florence.

Abstract: Background/Objectives: Parent–child relationships represent a key factor for the quality of developmental trajectories and impact on children's social and emotional competence. Therefore, research has advanced the role of parenting by showing the significance of differentiating between distinctive aspects of a parent's behaviors. This study aims to investigate the role of the feelings experienced in parent–child relationships (e.g., warmth and negative feelings), considering the moderating role of the parental styles toward children's emotions (e.g., coaching and dismissing). Methods: A total of 136 mothers (M = 38.09 years, SD = 4.51 anni, 48.5% high school degree) with a preschool child (age range 3–5 years) in Central Italy have been involved in a survey during the pandemic period. Results: Multiple regression analyses show that warmth and negative feelings are associated with positive parenting; however, the moderation effect of the dismissing style on both warmth and negative feelings emerged. Conclusions: Despite the characteristics of the data collection period, the results suggest the importance of considering the emotion-related dimensions between parents and their children as they seem to influence parenting behaviors.

Keywords: parenting; parental feelings; children's emotions; developmental trajectories

Citation: Facci, C.; Baroncelli, A.; Ciucci, E. The Role of Affects and Emotional Styles in the Relationship Between Parents and Preschool Children. *Children* **2024**, *11*, 1369. https://doi.org/10.3390/children11111369

Academic Editor: Matteo Alessio Chiappedi

Received: 11 October 2024
Revised: 4 November 2024
Accepted: 7 November 2024
Published: 12 November 2024

1. Introduction

Children's socio-emotional development relies on the interaction between individual and contextual aspects. Research suggests that children's emotional health could be linked to the association between aspects related to the child's characteristics (e.g., child temperament) and quality of parenting (e.g., positive and negative parenting) [1,2]. Particularly, research established that parents' behaviors and emotional competence—principally their emotion socialization behaviors and their emotional regulation—model the process of children's emotional socialization [3,4] and impact children's social and emotional competence as well as their maladaptive behaviors (e.g., conduct problems and CU behaviors) [5–7]. The ability to express, understand, and regulate emotions is a crucial element in socio-emotional development, particularly in young children. Moreover, the ability to identify and appropriately respond to emotional cues during a social interaction shapes their relationships. Research supports the idea that children's socio-emotional competence has its developmental roots in the preschool years, and patterns of parenting appear to play an important role during early childhood [8].

According to the developmental perspective, parenting has long been described considering parents' behaviors. Different levels of responsiveness and demandingness describe an authoritative, authoritarian, permissive, or indulgent style [9]. Recently, parenting has been recognized as a multidimensional construct, studied in terms of parent–child interactions that involve affection and acceptance (i.e., warmth) and discipline and rejection (i.e., control) [10]. The parent–child relationship has the power to shape the emotional climate of the family because it helps children feel supported and emotionally safe (e.g., free to express emotions) and is essential for the emotional socialization process and their socio-emotional development [11]. Parents have long been considered the first and most important socializers of emotional and social competence in the lives of their children [3,12]. The influences of the parent's responsiveness and behaviors on children's emotional socialization constitute a focal issue in developmental psychology [11,13]. Parent–child dyadic mutuality (i.e., shared positive affect, responsiveness, and cooperation) is recognized as an important component of family socialization processes [14,15]. Parents' discussion and expression of emotions or reactions to their child's emotions could outline the emotional climate of the family [3,16], shaping the quality of parent–child interactions. Also, parental beliefs and feelings about emotions drive parents' emotional socialization behaviors toward their children [17–19]. The parental emotional styles (i.e., coaching and dismissing), based on beliefs and feelings, in responses to children's negative emotions have been identified as a key feature for children's socio-emotional development, because of their relations with children's developing socio-emotional skills and difficulties [11,18].

1.1. Parents' Emotional Competence and Meta-Emotion Philosophy

The emotional competence of the parents, which can be defined as the multifaceted ability to be aware of one's own and others' emotions, is predictive of parent–child relationship quality and the child's behavioral outcomes [12]. Parent's own emotions have received increasing attention in developmental research considering the influence on parenting behaviors and their role in creating the affective environment in which the children have been raised, which has an impact on their emotional and behavioral adjustment [20,21]. In addition, the rigor of the evaluation of parents' emotional experiences is also given increased attention [22]. Parents with problems in accepting their own emotions may have difficulties engaging in supporting their children's emotional socialization behaviors. When parents experience high levels of negative emotions, they may feel overwhelmed or 'flooded', increasing the likelihood of withdrawal or suppression of negative emotions, resulting in dismissal, or punitive discipline [23,24]. Parental problems in recognizing, accepting, and regulating their emotions tend to decrease emotional expressiveness and supportive parenting [21,24]. When parents have impairments in emotional competence and are less accepting of their own emotions, they may be less likely to talk overall about feelings and emotions [25].

Gottman, Katz, and Hooven [17] have proposed that parents hold a meta-emotion philosophy that involves their thoughts and feelings about their own emotions and the emotions of their children, and this is connected to the process of how they socialize emotions with their children. In more recent years, Katz and colleagues [18] showed and updated the meta-emotion philosophy framework that sustains the importance of parental beliefs, feelings, and attitudes toward emotions in children's socio-emotional development. The belief that emotions are to be validated implies that there is recognition and acceptance of emotions as developmental benefits and opportunities for children to learn. Instead, the belief that emotions are dangerous may suggest less awareness and acceptance of emotions as an occasion to be supportive [13,26,27]. Parents' beliefs and feelings about emotion shape parents' perception of their child's emotional experiences and their thoughts about how to teach emotions to their children, which mainly results in two emotional styles: coaching and dismissing [16,18,26]. Parents who dismiss or disapprove of the expression of negative emotions (i.e., sadness and anger) teach their children that these emotions are problematic or dangerous and perform behaviors such as denying, ignoring, or minimizing children's emotions. Further, parents who coach the expression of negative emotions are supportive of their children's expressions of these emotions and adopt behaviors such as emotional scaffolding, praising, and validation [17,28]. Although emotion coaching and emotion dismissing may seem to be opposite emotional styles, observational studies with children in middle childhood suggest that parents who engaged in both coaching and dismissing of children's negative emotions had children with the lowest emotional dysregulation [29].

According to the updated meta-emotion framework of Katz and colleagues [18], parental meta-emotion philosophy, which is expressed by means of coaching and/or dismissing styles toward their children's emotions, could have an impact on parenting behaviors and be influenced by both parent and child emotional experiences. In a recent study, it has been suggested that parental emotions influenced parenting behaviors, but they did not necessarily determine them [30]; parents could be guided by their beliefs about emotions to engage in consistent parenting behaviors [19,27]. Parental emotional styles may be particularly sensitive to parents' emotional experiences because these styles reflect parents' meta-emotion philosophy within parent–child interactions [26], and, consequently, these styles reflect parents' intentions to scaffold or avoid children's emotional expression and exploration. Gottman and colleagues [17,26], already argued that parental emotional coaching style affects parents' inhibition of negative affect toward their children and facilitates positive parenting. The research on parents' meta-emotion philosophy has been generative, and the assessment of this construct has been improved [21,31]. This construct incorporates different aspects of parent–child interaction (e.g., emotional beliefs, parenting strategies, and children's emotional experiences) that require to be distinguished in order to better understand the process of socialization of emotion. Thus, an in-depth understanding of the role and relevance of parental meta-emotion philosophy, operationalized as parental coaching and dismissing styles, and the distinction between different parental aspects related to their emotional experience (i.e., parental feelings and behaviors) is essential when considering research on the emotion socialization process of children.

1.2. The Current Study

Based on the above-reported literature, the influence of the emotional experience of the parents in the parent–child relationship emerges. The main aim of the present study is to explore different dimensions of parenting, considering emotion-related dimensions (i.e., feelings and emotional style) as well as parental behaviors (i.e., positive, inconsistent, and punitive parenting) in a sample of mothers of preschool children.

In so doing, the associations between emotion coaching and emotion dismissing styles and parental feelings (i.e., warmth and negative feelings) in acting positive, inconsistent, and punitive parenting have been explored. Particularly, the aim was to examine whether parents may change their parental behaviors in association with parental feelings at different levels of emotional styles. A moderated model was tested to examine the unique and interactive effects between emotional styles and feelings in their association with parental behaviors. We hypothesized that parents experiencing negative emotions toward their children will be more likely to engage in negative parental behaviors at higher levels of beliefs that emotions can be problematic or dangerous (i.e., high level of dismissing style) [23,24]. We further hypothesized that parental coaching (i.e., recognition, acceptance, and regulation of children's negative emotions, considering them as an opportunity for closeness) will moderate the relationships between negative feelings and engaging in positive parenting behaviors [25,26].

2. Materials and Methods

2.1. Participants and Procedure

The sample (N = 163) was recruited from kindergartens in Central Italy during the period of October and November 2021. Participants were all mothers, and the children ranged in age from 3 to 5 years (56 children in the class of the "oldest"—5 years old; 35 in the class of the "medium"—4 years old; 68 in the class of the "youngest"—3 years old; 4 attending a mixed class), which for the Italian educational system is the age range at which children can attend kindergartens. The mothers ranged in age from 26 to 49 years (Mage = 38.09, Ds = 4.51, 48.5% high school degree); however, three mothers did not report their age. Across samples, most mothers were from Italian cultural backgrounds (91.87%). The three scholastic institutions involved in the research approved and collaborated with all procedures. Participants were presented with a description of the study that was developed in collaboration with scholastic institutions and teachers, followed by a request to complete an informed consent to participate. Mothers provided informed consent and then completed an internet survey about their child, with demographic information and details about the parents (e.g., school degree, family composition). A series of scales were presented in a Google Form survey session, with no time restriction to fill out the survey.

2.2. Measures

2.2.1. Parenting Behaviors

In the present study, we used the Alabama Parenting Questionnaire (APQ; [32]) preschool version (APQ-Pr; [33]) in the Italian version [34], which is a 32-item self-report test. The measure considers three dimensions of parenting: the 7-item inconsistent parenting subscale, which measures a lack of follow-through with discipline (e.g., "You threaten to punish your child and then do not actually punish him/her"); the 5-item punitive parenting subscale, which assesses how often a parent engages in corporal or harsh discipline (e.g., "You spank your child with your hand when he/she has done something wrong"); and the 12-item positive parenting subscale, which comprises items describing positive reinforcement and parental involvement (e.g., "You let your child know when he/she is doing a good job with something"). The respondent estimates the frequency of occurrence of each behavior on a 5-point Likert scale: 1 = never, 2 = almost never, 3 = sometimes, 4 = often, and 5 = always. The internal consistencies for the current sample were acceptable for the positive parenting scale ($\alpha = 0.76$) and modest for the inconsistent parenting scale ($\alpha = 0.67$) and punitive parenting scale ($\alpha = 0.60$).

2.2.2. Parental Feelings

In the present study, we used the Parent Feelings Questionnaire (PFQ; [35]), which is a 24-item measure that assesses both positive and negative parental feelings toward their children, and it is widely used as a measure of parental warmth in preschool samples [12]. The psychometric properties of the measure have been explored in other work [6]. The measure includes statements about feelings that are rated on a 5-point Likert-type scale, ranging from 1 (completely agree) to 5 (completely disagree). The measure includes two subscales: the parental warmth subscale ($\alpha = 0.60$), which comprises 15 items assessing positive parental feelings (e.g., "When I think about this child, it usually gives me warm feelings"); and the negativity subscale ($\alpha = 0.90$), which comprises 16 items assessing negative parental feelings (e.g., "Sometimes I am not happy about my relationship with this child").

2.2.3. Maternal Emotional Style

To assess the maternal emotional coaching (EC) and emotional dismissing (ED) styles, the Maternal Emotional Styles Questionnaire—MESQ, developed by Legacé Séguin and Coplan [36] in the Italian translation [37], was used. The psychometric properties of the measure have been explored in other work [6]. This instrument assesses the maternal emotional behaviors produced in response to children's emotional displays. The 14 items represented a combination of fear, anger, and sadness emotions across the two emotional styles. Mothers are asked to express their level of agreement with each of the 14 items on a 5-point Likert scale ranging from 1 (strongly disagree) to 5 (strongly agree). Seven items describe EC (score range: 7–35; sample item: "When my child is angry, I take some time to experience this feeling with him/her") and Cronbach's alpha was 0.66 for this subscale; seven items describe ED (score range: 7–35; sample item: "When my child is angry, my goal is to make him/her stop") and Cronbach's alpha for this subscale was 0.81.

2.3. Data Analyses

All the analyses were conducted using SPSS version 28 [38]. First, bivariate correlations were used to examine the relations between all the variables. Then, a moderated model that examines the unique and interactive effects between emotional styles and feelings in their association with parental behaviors was performed. Hierarchical regression analyses explored the contributions of maternal feelings about children (i.e., each subscale of the PFQ) along with the potential moderating role of maternal emotional styles (i.e., both the emerged factors of the MESQ) to specific parental behaviors (i.e., positive, inconsistent, and punitive parenting). Although the directionality of the associations between the variables cannot be established in a cross-sectional study, the application of a regression approach requires assumptions about which variables to consider as independent and which to consider as criterion variables. The present analyses assumed that maternal feelings could influence maternal behaviors, and this is moderated by maternal emotional styles. Consequently, measures of feelings and behaviors were entered along with all scales of MESQ in Step 1, followed by the interaction terms between the two scales of MESQ and the two scales of feelings in Step 2. The form of results indicating significant interactions was explored using post hoc probing procedures indicated by Holmbeck [39].

3. Results

The distribution of study variables and bivariate correlations (Pearson's r) among the study variables are provided in Table 1.

Table 1. Descriptive statistics and correlations (Pearson's r) among study variables.

Variable	M	SD	Skew.	Kurt.	1	2	3	4	5	6	7
1 APQ—Positive Parenting	4.53	0.35	−0.92	0.57	/	−.01	−.04	.33 ***	−.35 ***	.16 *	.20 *
2 APQ—Inconsistent Parenting	2.55	0.71	0.11	−0.31		/	.16 *	−.20 **	.22 **	−.10	.39 ***
3 APQ—Punitive Parenting	2.17	0.63	0.60	0.52			/	−.43 ***	.34 ***	−.11	.16 *
4 PFQ—Warmth	4.68	0.39	−1.90	4.02				/	−.34 ***	.28 ***	−.10
5 PFQ—Negativity	2.31	0.80	0.40	−0.61					/	−.10	−.18 *
6 MESQ—Coaching	4.43	0.58	−1.19	1.50						/	−.06
7 MESQ—Dismissing	3.30	0.82	0.03	−0.67							/

Notes. APQ: Alabama Parenting Questionnaire. PFQ: Parent Feelings Questionnaire, MESQ: Maternal Emotional Style Questionnaire. * $p < 0.05$, ** $p < 0.01$, *** $p < 0.001$.

In this community sample, the distribution of the scales did not deviate significantly from normality. As expected, there were significant correlations between the three scales of parental behaviors and warmth and negativity. The coaching style was significantly positively correlated with positive parenting ($r = 0.16$, $p < 0.05$) and warmth ($r = 0.28$, $p < 0.001$). In addition, the dismissing style was positively correlated with all the parenting behaviors and negatively with negativity ($r = -0.18$, $p < 0.05$).

The results of the multiple regression analyses, testing the unique and interactive effects between emotional styles and feelings in their association with parental behaviors, are reported in Table 2.

Multiple regression analyses showed that warmth ($\beta = 0.30$, $p < 0.01$) and negativity ($\beta = -0.32$, $p < 0.01$) are associated with positive parenting; however, the moderation effect of dismissing style on warmth ($\beta = -0.25$, $p < 0.01$) and negativity ($\beta = -0.17$, $p < 0.05$) emerged. The form of the interaction provided in Figure 1 indicates that higher levels of positive feelings were related to higher positive parenting with low dismissing style ($\beta = 0.59$, $p < 0.001$) but not with high dismissing style ($\beta = -0.02$, $p > 0.05$). The form of the interaction provided in Figure 2 indicates that higher levels of negativity were related to lower positive parenting with a high dismissing style ($\beta = -0.50$, $p < 0.001$) but not with a low dismissing style ($\beta = -0.14$, $p > 0.05$). Moreover, multiple regression analyses showed that negativity ($\beta = 0.29$, $p < 0.05$) and dismissing style ($\beta = 0.44$, $p < 0.001$) are associated with inconsistent parenting; the dismissing style emerged positively associated with inconsistent parenting ($\beta = 0.44$, $p < 0.001$), and no moderation effect emerged between feelings and maternal emotional styles ($F(10,162) = 5.256$; $p < 0.001$; $\Delta R^2 = 0.01$, $p = 0.88$, $R^2 = 0.21$). Moderate multiple regression analyses showed that warmth ($\beta = -0.34$, $p < 0.01$) and negativity ($\beta = 0.33$, $p < 0.001$) are associated with punitive parenting; the dismissing style emerged positively, even moderately, associated with punitive parenting ($\beta = 0.18$, $p < 0.05$), and no moderation effect emerged ($F(10,162) = 6.334$; $p < 0.001$; $\Delta R^2 = 0.005$, $p = 0.56$, $R^2 = 0.25$).

Table 2. Regression analyses testing the main and interactive effects of parental feelings and parental emotion socialization style on parenting behaviors.

	APQ-Positive	APQ-Inconsistent	APQ-Punitive	PFQ-Warmth	PFQ-Negativity	Coaching	Dismissing	F	R^2
APQ-Positive	-	.02	.16 *	.30 ** (a)	−.32 ** (b)	.11	.17 *	(10,162) = 6.788 ***	.26
APQ-Inconsistent	.02	-	−.05	−.10	.29 ***	−.07	.44 ***	(10,162) = 5.256 ***	.21
APQ-Punitive	.16 *	−.04	-	−.34 ***	.32 ***	−.04	.18 *	(10,162) = 6.334 ***	.25

Notes. APQ: Alabama Parenting Questionnaire. PFQ: Parent Feelings Questionnaire. * $p < 0.05$, ** $p < 0.01$, *** $p < 0.001$. (a) There was a significant two-way interaction effect with dismissing in the association between warmth and positive parenting ($\beta = -0.25$, $p < 0.01$; $F(10,162 = 6.788)$, $p < 0.001$; $\Delta R^2 = 0.05$, $p < 0.01$, $R^2 = 0.26$), indicating that higher levels of warmth were related to higher positive parenting with low dismissing style ($\beta = 0.59$, $p < 0.001$) but not with high dismissing style ($\beta = -0.02$, $p > 0.05$). (b) There was a significant two-way interaction effect with dismissing in the association between negativity and positive parenting ($\beta = -0.17$, $p < 0.05$, $F(10,162 = 6.788)$, $p < 0.001$, $\Delta R^2 = 0.05$, $p < 0.01$, $R^2 = 0.26$), indicating that higher levels of negativity were related to lower positive parenting with high dismissing style ($\beta = -0.50$, $p < 0.001$) but not with low dismissing style ($\beta = -0.14$, $p > 0.05$).

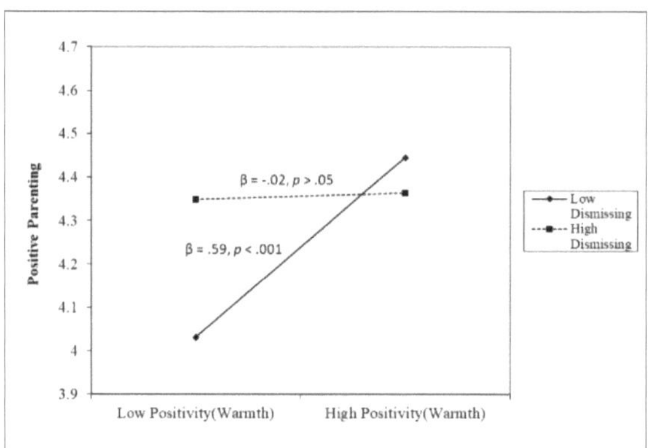

Figure 1. The moderating role of dismissing in the association between warmth and positive parenting.

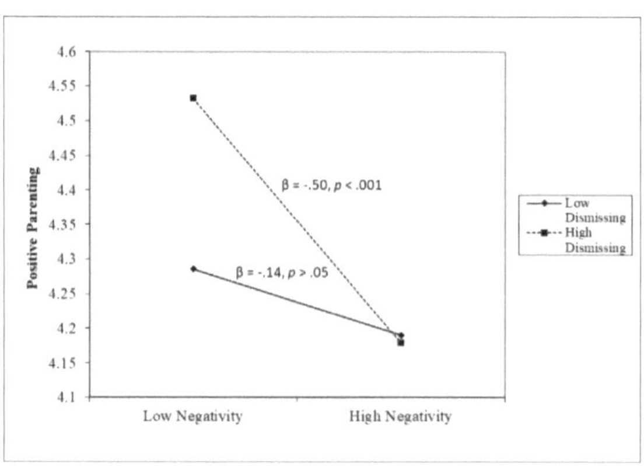

Figure 2. The moderating role of dismissing in the association between negativity and positive parenting.

4. Discussion

The current study investigated the role of emotion-related dimensions in engaging parental behaviors. Particularly, it explored the associations between parental emotional styles (e.g., coaching and dismissing styles) and feelings in acting parenting behaviors.

As expected, we found an association between the maternal emotional styles and the other parenting variables. Particularly, we found positive associations among warmth feelings, positive parenting scale, and coaching style. A negative association emerged between dismissing style and negative maternal feelings. We also found a positive association between the dismissing style and all the parenting behaviors (i.e., positive, inconsistent, and punitive parenting). Maternal warmth is strongly associated with parenting behaviors; positively with positive parenting but negatively with inconsistent and punitive parenting. Negative feelings are positively associated with inconsistent and punitive parenting but negatively associated with positive parenting. The result that maternal emotional style interacts with parenting behaviors and feelings is in line with the Parental Meta-Emotion Philosophy (PMEP) theory [18], which hypothesized that emotional socialization behaviors are guided by parents' own emotion-related beliefs and feelings. Particularly, parents with a coaching style are responsive to their children's emotions, support them to problem-solve, and are aware of the positive feelings about the relationship with the children. Surprisingly, a parent with a dismissing style who tends to reject or dismiss children's emotions expresses fewer negative feelings and performs both positive and negative parenting behaviors. It is known that parental behaviors can be influenced by differences in children's emotionality. Parents may experience distress and alter their views on how to teach their children about emotions [16–18] and how to relate based on their child's emotionality. Children expressing negative emotions may make it harder for parents to engage in supportive emotion socialization behaviors. A parent who engages in a dismissing style is a parent who is uncomfortable with the expression of emotions, above all the negative ones. It may be speculated that mothers higher in dismissing style consider the child's negative emotions as interpersonal challenges that put them out of their control and not as situations during which they perform their supportive parental role with their child. However, it has been shown that dismissing and coaching may not be the opposite style [29], and parents can adopt the dismissing style in the specific situation, engaging in both positive and negative parenting behaviors. There is evidence, even mixed, that parents' emotions influence parenting behaviors [3,11]. It is suggested that parents with more understanding of their emotions are better able to apply supportive parenting behaviors [21,24]; various feelings may make it harder for parents to engage in positive parenting and a supportive emotional style [4]. Moreover, recognition and acceptance of negative feelings could be a protective factor from engaging in negative parenting behaviors [20].

A moderated model was tested to examine the unique and interactive effects between emotional styles and feelings in their association with parenting behaviors. We combined maternal feelings and maternal emotional styles into a regression model to highlight their role in parenting behaviors. The tested moderation model led us to make an in-depth exploration of the connection between the emotion-related parental dimensions. We expected to find the moderating role of both emotional styles on the association between negative feelings and parenting behaviors [23,24,28]. Contrary to our hypotheses, we found the moderating role of parental emotional styles only on positive parenting behaviors but not on negative ones. Our results suggested only the moderating role of maternal emotion dismissing style on the relation between both warmth and negative feelings on positive parenting; higher levels of warmth were related to higher positive parenting with low dismissing style. Warmth feelings could increase the probability of implementing positive parenting in parents with a low emotion dismissing approach to children's emotions. Higher levels of negative feelings were related to lower positive parenting with a high-dismissing style. Negative feelings seemed to reduce the probability of implementing positive parenting in mothers with a high emotion dismissing style. It appears that warmth feelings are a protective factor for mothers with a low emotion dismissing style, and negative feelings are a risk factor for mothers with a high dismissing style. Mothers who believe that emotions can be problematic or dangerous for children (i.e., high emotional dismissing style) may hide or mask their own emotions in attempts to shield children from observing their emotional experiences. Therefore, the effort to dismiss emotions could create difficulties in engaging in positive parenting. At the same time, mothers who do not value emotions as challenging or risky (low dismissing style) may thus be more supportive and emotionally accepting toward the child's emotions and experience positive feelings toward the child. We may suppose that emotion-dismissing parents are not less effectively involved than emotion-coaching parents; they may simply base their behaviors on an attitude that minimizes the role of feelings. Furthermore, parents may react positively or negatively to certain emotions depending on the nature of the emotion (e.g., sadness vs. anger). Thus, rather than simply sticking to a set of beliefs around emotions, parents may choose to behave with involvement and responsiveness in ways that are situationally and personally relevant to them and to their children.

Our results should be read in the context of some limitations. First, the cross-sectional nature of this study prevents us from making causal inferences. Future research should use longitudinal designs to explore the dynamics or causal relations among these variables. Second, both the size and representativeness of our sample limit the generalizability of results. The sample used in this study was primarily Italian, from central regions, and was of all mothers; these characteristics limit results interpretation to specific cultural and gender backgrounds. Different cultural perspectives may affect how parents think about emotions, specifically about children's emotions [2,13]. Moreover, some studies have gone beyond the mother-child dyad and explored the importance of father-child dynamics; future research should explore the emotional style of fathers that could influence in different ways the children's adjustments [2,19,40]. Another important aspect that could be considered in future research could be the income of the families that may be associated with a risk factor, such as stress, that can affect the quality and quantity of parental emotion styles [41]. Third, the data were collected only through parent report measures, and this may affect the risk of biasing the subjectivity of the results. Thus, future analyses could combine different measures, such as observational measures of the parent–child interactions.

5. Conclusions

The results of this study suggest the importance of considering the emotional dimensions of the parents toward their children, both in terms of feelings and emotional styles, as they would seem to influence parenting behaviors. Overall, our results suggest that during childhood, the parents' belief about negative emotions to be denied influences their parenting positive behaviors, experiencing both positive and negative feelings. Parents controlling their negative feelings could make it harder to engage in positive parenting, but positive feelings protect them from not being supportive of the negative emotions of their children. Results from the present study should be interpreted considering the period of data collection (October–November 2020). The pandemic period could have influenced the family dynamics since a large body of research has identified a negative impact on individuals' emotional health, both adults and children [42,43].

In conclusion, parents who are responsive and warm typically display specific types of parenting behaviors and have certain beliefs associated with emotions that lead to coaching and dismissing styles [3,25,26]. Research needs to examine how different parent–child interactions, particularly emotion-related dimensions, act together to outline parents' behaviors. The results should be used to implement educational programs or training interventions for parents. Such programs should encourage both the level of emotional expressiveness and parents' own emotion regulation skills [44].

Author Contributions: Conceptualization, C.F., A.B. and E.C.; methodology, C.F., A.B. and E.C.; formal analysis C.F., A.B. and E.C.; data curation, C.F. and E.C.; writing—original draft preparation, C.F. and E.C.; writing—review and editing, C.F., A.B. and E.C. All authors have read and agreed to the published version of the manuscript.

Funding: This research received no external funding.

Institutional Review Board Statement: The study was conducted in accordance with the Declaration of Helsinki, in accordance with the recommendations of the Italian Association of Psychology, and approved by the Ethical Commission of the University of Florence (protocol code n. 142—26 February 2021).

Informed Consent Statement: Informed consent was obtained from all subjects involved in the study.

Data Availability Statement: There are no unpublished data available. The corresponding author can be contacted regarding this matter.

Acknowledgments: This research is part of Facci's PhD thesis in Education Sciences and Psychology at the University of Florence with the title *"Parenting and Parental Emotional Styles: Which relations with parental affects towards children and children's callous-unemotional behaviors?"* The authors are grateful to the Scholastic Institutions that supported the project by agreeing to conduct research at their schools. Moreover, the authors are thankful to the families who participated in the project.

Conflicts of Interest: The authors declare no conflicts of interest.

References

1. Danzig, A.P.; Dyson, M.W.; Olino, T.M.; Laptook, R.S.; Klein, D.N. Positive Parenting Interacts with Child Temperament and Negative Parenting to Predict Children's Socially Appropriate Behavior. *J. Soc. Clin. Psychol.* **2015**, *34*, 411–435. [CrossRef] [PubMed]
2. Rothenberg, W.A.; Ali, S.; Rohner, R.P.; Lansford, J.E.; Britner, P.A.; Di Giunta, L.; Dodge, K.A.; Malone, P.S.; Oburu, P.; Pastorelli, C.; et al. Effects of Parental Acceptance-Rejection on Children's Internalizing and Externalizing Behaviors: A Longitudinal, Multicultural Study. *J. Child Fam. Stud.* **2022**, *31*, 29–47. [CrossRef]
3. Morris, A.S.; Silk, J.S.; Steinberg, L.; Myers, S.S.; Robinson, L.R. The Role of the Family Context in the Development of Emotion Regulation. *Soc. Dev.* **2007**, *16*, 361–388. [CrossRef] [PubMed]
4. Rutherford, H.J.; Wallace, N.S.; Laurent, H.K.; Mayes, L.C. Emotion Regulation in Parenthood. *Dev. Rev.* **2015**, *36*, 1–14. [CrossRef] [PubMed]
5. Duncombe, M.E.; Havighurst, S.S.; Holland, K.A.; Frankling, E.J. The contribution of parenting practices and parent emotion factors in children at risk for disruptive behavior disorders. *Child Psychiatry Hum. Dev.* **2012**, *43*, 715–733. [CrossRef] [PubMed]
6. Facci, C.; Baroncelli, A.; Frick, P.J.; Ciucci, E. Dimensions of Parenting and Children's Conduct Problems: The Importance of Considering Children's Callous-Unemotional Traits. *Int. J. Environ. Res. Public Health* **2024**, *21*, 779. [CrossRef]
7. Kochanska, G.; Boldt, L.J.; Kim, S.; Yoon, J.E.; Philibert, R.A. Developmental interplay between children's biobehavioral risk and the parenting environment from toddler to early school age: Prediction of socialization outcomes in preadolescence. *Dev. Psychopathol.* **2015**, *27*, 775–790. [CrossRef]
8. Kochanska, G.; Boldt, L.J.; Goffin, K.C. Early relational experience: A foundation for the unfolding dynamics of parent-child socialization. *Child Dev. Perspect.* **2019**, *13*, 41–47. [CrossRef]
9. Maccoby, E.E.; Martin, J.A. Socialization in the context of the family: Parent-child interaction. In *Handbook of Child Psychology: Vol. IV. Socialization, Personality, and Social Development*; Mussen, P., Hetherington, E., Eds.; Wiley: New York, NY, USA, 1983; pp. 1–101.
10. Deater-Deckard, K.; Lansford, J.E.; Malone, P.S.; Alampay, L.P.; Sorbring, E.; Bacchini, D.; Bombi, A.S.; Bornstein, M.H.; Chang, L.; Di Giunta, L.; et al. The association between parental warmth and control in thirteen cultural groups. *J. Fam. Psychol.* **2011**, *25*, 790–794. [CrossRef]
11. Johnson, A.M.; Hawes, D.J.; Eisenberg, N.; Kohlhoff, J.; Dudeney, J. Emotion socialization and child conduct problems: A comprehensive review and meta-analysis. *Clin. Psychol. Rev.* **2017**, *54*, 65–80. [CrossRef]
12. Denham, S.A.; Wyatt, T.M.; Bassett, H.H.; Echeverria, D.; Knox, S.S. Assessing social-emotional development in children from a longitudinal perspective. *J. Epidemiol. Community Health* **2009**, *63* (Suppl. 1), i37–i52. [CrossRef] [PubMed]
13. Parker, A.E.; Halberstadt, A.G.; Dunsmore, J.C.; Townley, G.; Bryant, A.; Thompson, J.A.; Beale, K.S. Emotions are a window into one's heart": A qualitative analysis of parental beliefs about children's emotions across three ethnic groups. *Monogr. Soc. Res. Child. Dev.* **2012**, *77*, 1–136. [CrossRef] [PubMed]
14. Kim, S.; Boldt, L.J.; Kochanska, G. From parent-child mutuality to security to socialization outcomes: Developmental cascade toward positive adaptation in preadolescence. *Attach. Hum. Dev.* **2015**, *17*, 472–491. [CrossRef] [PubMed]
15. Deater-Deckard, K.; O'Connor, T.G. Parent-child mutuality in early childhood: Two behavioral genetic studies. *Dev. Psychol.* **2000**, *36*, 561–570. [CrossRef] [PubMed]
16. Eisenberg, N.; Cumberland, A.; Spinrad, T.L. Parental Socialization of Emotion. *Psychol. Inq.* **1998**, *9*, 241–273. [CrossRef]
17. Gottman, J.M.; Katz, L.F.; Hooven, C. *Meta-Emotion: How Families Communicate Emotionally*, 1st ed.; Routledge: Mahwah, NJ, USA, 1997. [CrossRef]

18. Katz, L.F.; Maliken, A.C.; Stettler, N.M. Parental meta—Emotion philosophy: A review of research and theoretical framework. *Child Dev. Perspect.* **2012**, *6*, 417–422. [CrossRef]
19. Dunsmore, J.C.; Her, P.; Halberstadt, A.G.; Perez-Rivera, M.B. Parents' Beliefs about Emotions and Children's Recognition of Parents' Emotions. *J. Nonverbal Behav.* **2009**, *33*, 121–140. [CrossRef]
20. Dunsmore, J.C.; Booker, J.A.; Ollendick, T.H. Parental Emotion Coaching and Child Emotion Regulation as Protective Factors for Children with Oppositional Defiant Disorder. *Soc. Dev.* **2013**, *22*, 444–466. [CrossRef]
21. Meyer, S.; Raikes, H.A.; Virmani, E.A.; Waters, S.; Thompson, R.A. Parent Emotion Representations and the Socialization of Emotion Regulation in the Family. *Int. J. Behav. Dev.* **2014**, *38*, 164–173. [CrossRef]
22. Zhang, X.; Beatty, A.; Abela, K.; Melo, M.F.; Kenny, M.; Atkinson, L.; Gonzalez, A. Assessing parental emotion regulation in the context of parenting: A systematic review. *Dev. Rev.* **2023**, *69*, 101092. [CrossRef]
23. Lorber, M.F.; Mitnick, D.M.; Slep, A.M. Parents' experience of flooding in discipline encounters: Associations with discipline and interplay with related factors. *J. Fam. Psychol.* **2016**, *30*, 470–479. [CrossRef] [PubMed]
24. Mence, M.; Hawes, D.J.; Wedgwood, L.; Morgan, S.; Barnett, B.; Kohlhoff, J.; Hunt, C. Emotional flooding and hostile discipline in the families of toddlers with disruptive behavior problems. *J. Fam. Psychol.* **2014**, *28*, 12–21. [CrossRef] [PubMed]
25. Halberstadt, A.G.; Eaton, K.L. A Meta-Analysis of Family Expressiveness and Children's Emotion Expressiveness and Understanding. *Marriage Fam. Rev.* **2002**, *34*, 35–62. [CrossRef]
26. Gottman, J.M.; Katz, L.F.; Hooven, C. Parental meta-emotion philosophy and the emotional life of families: Theoretical models and preliminary data. *J. Fam. Psychol.* **1996**, *10*, 243–268. [CrossRef]
27. Lozada, F.T.; Halberstadt, A.G.; Craig, A.B.; Dennis, P.A.; Dunsmore, J.C. Parents' Beliefs about Children's Emotions and Parents' Emotion-Related Conversations with Their Children. *J. Child Fam. Stud.* **2016**, *25*, 1525–1538. [CrossRef]
28. Cleary, R.; Katz, L.F. Family-level emotion socialization and children's comfort with emotional expressivity. *Fam. Psychol.* **2008**, *24*, 6–13.
29. Lunkenheimer, E.S.; Shields, A.M.; Cortina, K.S. Parental emotion coaching and dismissing in family interaction. *Soc. Dev.* **2007**, *16*, 232–248. [CrossRef]
30. Hajal, N.J.; Paley, B. Parental emotion and emotion regulation: A critical target of study for research and intervention to promote child emotion socialization. *Dev. Psychol.* **2020**, *56*, 403–417. [CrossRef]
31. Paterson, A.D.; Babb, K.A.; Camodeca, A.; Goodwin, J.; Hakim-Larson, J.; Voelker, S.; Gragg, M. Emotion-Related Parenting Styles (ERPS): A Short Form for Measuring Parental Meta-Emotion Philosophy. *Early Educ. Dev.* **2012**, *23*, 583–602. [CrossRef]
32. Shelton, K.K.; Frick, P.J.; Wootton, J. Assessment of parenting practices in families of elementary school-age children. *J. Clin. Child Psychol.* **1996**, *25*, 317–329. [CrossRef]
33. Clerkin, S.M.; Marks, D.J.; Policaro, K.L.; Halperin, J.M. Psychometric properties of the Alabama parenting questionnaire-preschool revision. *J. Clin. Child Adolesc. Psychol.* **2007**, *36*, 19–28. [CrossRef] [PubMed]
34. Benedetto, L.; Ingrassia, E.M. L'Alabama Parenting Questionnaire per la fascia prescolare (APQ-Pr). Contributo all'adattamento italiano. *Disturbi Attenzione Iperattività* **2014**, *9*, 151–167.
35. Deater-Deckard, K. Parent Feelings Questionnaire. *J. Abnorm. Psychol.* **1996**. [CrossRef]
36. Lagaceé-Séguin, D.G.; Coplan, R.J. Maternal emotional styles and child social adjustment: Assessment, correlates, outcomes and goodness of fit in early childhood. *Soc. Dev.* **2005**, *14*, 613–636. [CrossRef]
37. Ciucci, E.; Menesini, E. La comprensione delle emozioni nei bambini e lo stile emotivo materno. In *Lo Sviluppo Della Comprensione Delle Emozioni e la sua Valutazione*; Albanese, O., Molina, P., Eds.; Unicopli: Milano, Italy, 2008; pp. 201–208.
38. IBM Corp. *IBM SPSS Statistics for Windows, Version 28.0.*; IBM Corp: Armonk, NY, USA, 2021.
39. Holmbeck, G.N. Post-hoc Probing of Significant Moderational and Mediational Effects in Studies of Pediatric Populations. *J. Pediatr. Psychol.* **2002**, *27*, 87–96. [CrossRef]
40. Baker, J.K.; Fenning, R.M.; Crnic, K.A. Emotion Socialization by Mothers and Fathers: Coherence among Behaviors and Associations with Parent Attitudes and Children's Social Competence. *Soc. Dev.* **2011**, *20*, 412–430. [CrossRef]
41. Taraban, L.; Shaw, D.S. Parenting in context: Revisiting Belsky's classic process of parenting model in early childhood. *Dev. Rev.* **2018**, *48*, 55–81. [CrossRef]
42. Matiz, A.; Fabbro, F.; Paschetto, A.; Urgesi, C.; Ciucci, E.; Baroncelli, A.; Crescentini, C. The Impact of the COVID-19 Pandemic on Affect, Fear, and Personality of Primary School Children Measured During the Second Wave of Infections in 2020. *Front. Psychiatry* **2022**, *12*, 803270. [CrossRef]
43. Facci, C.; Iacopino, M.; Baroncelli, A.; Ciucci, E. The rise of online teaching and digital learning during the health emergency from COVID-19 and teachers' working self-efficacy: An Italian perspective. In Proceedings of the 3rd Symposium on Psychology-Based Technologies, Naples, Italy, 4–5 October 2021; pp. 1–7. Available online: https://ceur-ws.org/Vol-3100/paper10.pdf (accessed on 1 March 2024).
44. Havighurst, S.S.; Wilson, K.; Harley, A.E.; Prior, M.R.; Kehoe, C. Tuning in to kids: Improving emotion socialization practices in parents of preschool children—Findings from a community trial. *J. Child. Psychol. Psychiatry* **2010**, *51*, 1342–1350. [CrossRef]

children MDPI

Article

Infant Crying, Sleeping, and Feeding Problems in Times of Societal Crises: The Mediating Role of Parenting Stress on Parenting Behavior in Fathers and Mothers

Katharina Richter [1,2,†], Anna Friedmann [1,2,*,†], Volker Mall [1,2,3] and Michaela Augustin [1]

1 School of Medicine and Health, Social Pediatrics, Technical University of Munich, 81675 Munich, Germany; katharina.richter@tum.de (K.R.); volker.mall@tum.de (V.M.); michaela.augustin@tum.de (M.A.)
2 German Center for Child and Adolescent Health (DZKJ), Partner Site Munich, 80337 Munich, Germany
3 kbo-Kinderzentrum, 81377 Munich, Germany
* Correspondence: anna.friedmann@tum.de
† These authors contributed equally to this work.

Abstract: Background/Objectives: Infant regulatory problems (RPs), i.e., crying, sleeping, and feeding problems, are associated with unfavorable outcomes in later childhood. RPs increased during the pandemic; however, their occurrence in the face of today's societal challenges remains unclear. RPs are strongly linked to parenting stress and less positive parenting behaviors, but their interplay is less investigated. Methods: In this cross-sectional, questionnaire-based study (ntotal = 7039), we compared the incidences of crying, sleeping, and feeding problems in infants (0–2 years) in pandemic (npandemic = 1391) versus post-pandemic (npost-pandemic = 5648) samples in Germany. We also investigated the relationship between post-pandemic infant RPs and parenting behaviors with parenting stress as a potential mediator for fathers and mothers. Results: Crying/whining/sleeping problems (34.8%) and excessive crying (6.3%) were significantly more prevalent in the post-pandemic sample. In both mothers and fathers, infant RPs were significantly associated with less positive parenting behaviors. Parenting stress partially mediated this relationship. Conclusions: RPs in the post-pandemic era are even more prevalent than during the pandemic, highlighting the imperative for health care professionals to focus on infant mental health. Parenting stress emerges as an entry point for addressing the cycle of infant RPs and maladaptive behaviors in both fathers and mothers.

Keywords: infant mental health; stress; sensitivity; responsivity; over-reactivity; pandemic; regulatory problems

Citation: Richter, K.; Friedmann, A.; Mall, V.; Augustin, M. Infant Crying, Sleeping, and Feeding Problems in Times of Societal Crises: The Mediating Role of Parenting Stress on Parenting Behavior in Fathers and Mothers. *Children* **2024**, *11*, 1540. https://doi.org/10.3390/children11121540

Academic Editor: Matteo Alessio Chiappedi

Received: 18 November 2024
Revised: 10 December 2024
Accepted: 16 December 2024
Published: 19 December 2024

1. Introduction

Over the past few decades, longitudinal research has consistently demonstrated the harmful effects of early psychosocial stress on children's mental health, with repercussions that can persist throughout the lifespan [1,2]. Psychosocial stress in families can arise from a variety of sources, including challenging living conditions (e.g., low socioeconomic status, lack of social support, or isolation), strained family relationships (e.g., between parents and children), and specific family burdens (e.g., health issues or limited childcare resources) [3]. The COVID-19 pandemic exacerbated these stressors, with families facing a significant increase in psychosocial pressure [4,5].

Concerns regarding child mental health emerged early in the pandemic, with national and international studies showing a marked rise in psychological problems among children and adolescents compared to pre-pandemic levels [6,7]. Very young children are known to be particularly vulnerable to the harmful effects of environmental stress [8], as they are almost entirely reliant on their parents for emotional and physical support during a critical phase of rapid brain development [9,10] in which key milestones in emotional regulation and learning have to be met [11]. When infants struggle to adapt to new conditions, it can manifest in regulatory problems, i.e., crying, sleeping, and feeding problems [12]. Their prevalence varies significantly depending on the diagnostic criteria and assessment tools used, but studies suggest that between 3% and 40% of physically healthy children experience these behavioral difficulties during infancy and early childhood [13–15]. Although regulatory problems may resolve within a few months, they can persist in the form of various behavioral irregularities into the preschool and elementary school years [16,17]. The persistence of regulatory issues increases the risk for social–emotional and cognitive difficulties. Children with multiple regulatory problems are particularly at risk for both internalizing and externalizing disorders later in life [18]. Hence, addressing these problems as early as possible is crucial.

The available studies investigating infant regulatory problems in the context of the pandemic predominantly suggest at least partially elevated levels of these difficulties compared to pre-pandemic periods [19,20]. Findings indicate that maternal pre- and postnatal concerns related to the pandemic, alongside anxiety and reduced wellbeing resulting from pandemic-related restrictions, may be associated with increases in infant regulatory problems [21–25]. The CoronabaBY study, which examined psychosocial stress factors such as parenting stress and mental health problems in families with children aged 0–3 years in Southern Germany [25], revealed an increase in infant mental health problems during the course of the pandemic. Surprisingly, infant crying, sleeping, and feeding problems did not follow a linear pattern in response to societal restrictions but remained at a distinctively high level compared to pre-pandemic data even after the reduction in pandemic restrictions. The findings suggest a staggered negative impact of pandemic-related factors on young children's mental health that may still be present today, even if the pandemic is fading from public attention. This influence most likely operates indirectly through the wellbeing and mental health of the parents, which, in turn, shapes the emotional and psychological development of their infants. There is evidence from the pandemic period that psychological distress grows with the number of stressors present [26]. Since the aftermath of the pandemic, new stressors like, e.g., inflation, the climate crisis, and the ongoing wars in Ukraine and the Middle East, have emerged and, hence, families have had no time to rest navigating from one societal crisis to the next [27–29]. While societal crises might differ as to how they impact families (e.g., the pandemic having possibly combined direct and indirect effects and the recent wars probably having a more indirect influence in countries not involved in the conflict), it is evident that they occur as stress factors for families. This might be particularly critical in families with young children, as the transition to parenthood is already a challenging process. Family roles are often not yet established, and routines are not yet in place [30]. This may lead to an accumulation of family-related and societal stress factors in early parenthood with potential extensive impacts on infant mental health. Parents with their young children have indeed been identified as a particularly vulnerable group for mental health risks in this context. However, up to date, it remains largely unclear whether infant mental health problems in the post-pandemic period are as pronounced as they were during COVD-19.

Regulatory problems are usually demanding for parents, often leading to frustration, feelings of overwhelm, and negative emotions towards both the child and the parental role, which can strain the parent–child relationship [12]. Parenting stress and non-functional parenting behavior might further reinforce infant crying, sleeping, and feeding problems and could therefore be the relevant counseling entry points to break the cycle of negative family dynamics [31].

A shaping factor for the parent–child relationship is parenting behavior. Positive parenting behaviors, i.e., parental sensitivity and responsiveness, characterized by timely and appropriate responses to the child's emotional and physical cues, are crucial for child mental health, particularly in early childhood [32,33]. Sensitive and responsive parenting behaviors promote positive interactions with the child, which, in turn, encourage positive emotions and cognitions about the parenting role [32,34]. However, in the context of infant regulatory problems, a parent's ability to maintain this sensitivity and responsiveness can be impaired, making it harder to perceive and address the child's needs effectively [11]. Parents of children with crying, sleeping, or feeding problems may become more insecure towards their children's emotional signals, leading to more inadequate parenting behaviors, such as over-reactivity (exaggerated emotional responses or impulsive reactions to a child's behavior) [35].

One key factor associated with less positive parenting behaviors is parenting stress, i.e., the burden resulting from the demands of the parenting role [36]. As caring for children with crying, sleeping, and feeding problems is especially demanding, parenting stress is also known to be particularly pronounced in families encountering these infant mental health issues [37,38].

Taken together, regulatory problems are known to involve a complex interplay of infant behavioral problems in at least one developmental area, pronounced parenting stress, and dysfunctional interaction patterns when dealing with the infant's behavior [11,12]. Pandemic-based research has shown that parent–child relationships suffered [39], less positive parenting behaviors emerged [40], and parenting stress increased steadily over the course of COVID-19 [41]. First insights even showed a further increase in parenting stress following the declared end of the pandemic [42]. These results seem highly relevant in the context of regulatory problems today; however, their interplay is not yet clear. To the best of our knowledge, few studies have examined these variables within one integrated model, and research on infants with regulatory problems—such as problems with crying, sleeping, and feeding—remains notably limited in this context. Given the practical and clinical significance of these early regulatory difficulties, it is advisable for future research to place greater emphasis on infant crying, sleeping, and feeding problems within such models.

The existing literature on infant mental health is overwhelmingly based on mothers as caregivers. However, a broader societal trend has been observed over time, with an increasing expectation and willingness among fathers to take on a more active role in childrearing [43], reflected also by the rising number of fathers taking parental leave [44]. This shift has particularly intensified since the pandemic; due to working from home and limited access to childcare, fathers spent more time with their children and became more actively involved in daily family life and its challenges [45–47]. Some of these changes, such as remote working, have persisted post-pandemic. However, as most mothers still bear the majority of the caregiving responsibilities [48], fathers have often been overlooked in the research up to this point, negating their evolving roles in the family system.

The aim of this research is to deepen our understanding of the occurrence of infant crying, sleeping, and feeding problems in the post-pandemic context, as well as the dynamic with parenting stress and parenting behavior considering not only mothers but also the so-far-understudied group of fathers of infants with mental health problems. A thorough understanding will enable the development of more effective support strategies tailored to the specific needs of families, ensuring that both children and parents receive the appropriate care and intervention to promote healthy development and wellbeing also in times of societal crises.

Against this background, we aimed to answer the following research questions:

Are infant mental health problems in the so-called 'post-pandemic era' as pronounced as they were during the pandemic? To answer the first research question, we cross-sectionally compared the incidences of crying, sleeping, and feeding problems in a sample of families with children aged 0–2 years old collected during the pandemic with a sample surveyed after the pandemic.

Are infant crying, sleeping, and feeding problems associated with fathers' and mothers' parenting behaviors, more precisely, their sensitivity, responsivity, and over-reactivity? Does parenting stress mediate this relationship? To answer the second research question, we investigated fathers and mothers in the post-pandemic group.

2. Materials and Methods

This study is a cross-sectional comparison between a sample of parents with infants between 0 and 2 years surveyed during the pandemic and a sample investigated post-pandemic. We drew the samples from two datasets: a total of 2940 parents participated in the CoronabaBY study conducted between February 2021 and March 2022 in Bavaria (Southern Germany). Of these, 1391 (47.31%) were mothers and fathers of infants aged 0 to 2 years, who formed the pandemic sample for the present investigation. For further details on the CoronabaBY study, see [25,41] or [49]. The post-pandemic sample was drawn from the JuFaBY (Junge Familien in Bayern: Young Families in Bavaria) study, which is an ongoing investigation of psychosocial stress in families with young children (up to primary school age) in Bavaria. In this project, 20,807 caregivers participated between February 2023 and August 2024, of whom 5648 (27.14%) were parents of infants between 0 and 2 years old and thus were part of the post-pandemic sample. The study sample comprised a total of 7039 participants. Study protocols for both the CoronabaBY and JuFaBY projects were approved by the Ethics Committee of the Technical University of Munich (CoronabaBY: vote no. 322/20 S; JuFaBY: vote no. 2022-483_1-S-KH).

2.1. Participants and Procedure

All participants were recruited through the smartphone app 'My Pediatric Practice' (www.monks-aerzte-im-netz.de) (accessed on 10 November 2024) and were surveyed using a digital questionnaire. The app is a widely used tool in pediatric practice aimed at facilitating communication between pediatricians and parents in outpatient pediatric care. An invitation to participate in this study was sent to all app users with children in the relevant age group. Parents were informed about the project via a detailed information text and could digitally consent to participate if interested. The invitations to participate in the survey were linked to the regular check-ups conducted as part of pediatric preventive care. These check-ups occur in Germany at the ages of three weeks, three months, six months, twelve months, and then annually until school entry. About 97% of parents participate in these examinations [50]. Only parents who could not understand the project information due to a lack of German language skills were excluded from participation. After giving informed consent, the app-using parent filled out digital questionnaires. As an incentive and a token of appreciation, families received a small gift for their children. For further details, refer to publications such as [25,41].

Data collection was based on standardized questionnaires, which were presented online in the 'My Pediatric Practice' app and could be completed and submitted by participating parents. In addition to infant crying, sleeping, and feeding problems, parenting stress, parenting behavior (sensitivity, responsivity, over-reactivity), as well as sociodemographic characteristics were recorded.

2.2. Measurements

2.2.1. Sociodemographic Characteristics

General demographic characteristics of the parents and their children were surveyed, including sex and age, the caregiving situation (single parent/no single parent), information on nationality and highest educational attainment, and the number of children in the household.

2.2.2. Infant Crying, Sleeping, and Feeding Problems

On the 'Questionnaire for Crying, Sleeping, and Feeding' [51], the two subscales 'Crying/Whining/Sleeping' and 'Feeding', as well as the scale for 'Overall Infant Regulatory Problems', were used. Parents answered 38 questions about their infants' behaviors. The answers were based on four-point Likert scales ranging from 'never' to 'always'. Based on validated cut-off values, dichotomous outcomes for the areas 'Crying/Whining/Sleeping' (cut-off: 1.84; sensitivity: 87%; specificity: 92%) and 'Feeding' (cut-off: 1.27; sensitivity: 57%; specificity: 77%) were calculated as 'problematic' or 'non-problematic'. Furthermore, a total score for 'Infant Overall Regulatory Problems' was recorded based on raw values. The questionnaire can also provide indications of excessive crying according to the Wessel criterion ('Rule of Threes': crying for three or more hours a day, on more than three days a week, for more than three weeks) [52]. The validity of the questionnaire was confirmed by the high internal consistencies of the scales and correlations with behavioral diaries and clinical diagnoses [51].

2.2.3. Parenting Stress

To assess parenting stress, the parent domain of the German adaptation of the 'Parenting Stress Index (PSI)' ('Eltern-Belastungs-Inventar' (EBI)) [53] was used. High scores indicate limited parental resources for upbringing and caring for the child. The parent domain includes the following subscales: 'health' (parental health impairment as a cause or result of parenting stress), 'isolation' (lacking integration in social networks), 'role restriction' (perceived limitations as a result of being a parent), 'parental competence' (parental doubt about their own ability to manage upbringing and care for their child), 'attachment' (emotional relation of the parent to the child), 'depression' (worries and negative thoughts due to the parental role), and 'spouse-related stress' (as a result of being a parent). Answers were given on a 5-point Likert scale ranging from 1 = strongly agree to 5 = strongly disagree, resulting in a possible score range of 28–140, which can be converted into EBI Total Score T-values. The three cut-off categories for each subscale and the whole parent domain were 'not stressed' (T-value < 60), 'stressed' (T-value = 60–69), and 'strongly stressed' (T-value $\geq$ 70). The internal consistency of the parent domain was proven to be good (α = 0.93), and the retest reliability after one year was shown to be r = 0.87. Correlations with stress indicators and related constructs resulted in the assumption of the test validity [54].

2.2.4. Parenting Behavior

The 'Comprehensive Early Childhood Parenting Questionnaire (CECPAQ)' assesses parental behavior in early and middle childhood [55], drawing on attachment theory and social learning theories. The original questionnaire consists of 54 items and encompasses five core dimensions of parental behavior: 'parental support', 'stimulation', 'structuring', 'positive discipline', and 'harsh discipline'. Each dimension is further divided into two to three subscales, allowing for analysis at both the dimension and subscale levels. The study focused on three selected subscales based on suitability for the age group of 0–2 year olds. The subscales for 'responsiveness' and 'sensitivity' are part of the 'parental support' dimension, while the subscale for parental 'over-reactivity' belongs to the 'structuring' dimension. Responses to the questionnaire are measured on a six-point scale, ranging from 'never' to 'always'. Cronbach's alpha of 0.75 for mothers and 0.77 for fathers in the structuring scale, and 0.88 for both parents in the support scale were shown to be good/acceptable. Significant correlations with parenting stress and child behavioral problems indicate the high validity of the questionnaire [55].

2.2.5. Statistical Analysis

The data collected were initially provided to the research team by the operators of the 'My Pediatric Practice' app in the form of an Excel file for further processing. The datasets were then imported into the statistical software IBM SPSS Statistics Data Editor 28.0 for Windows. Since participants were required to submit fully completed questionnaires, there were very few missing data points. Individual missing data were excluded from the analysis.

To address the first research question, we tested whether infant crying, sleeping, and feeding problems differed between the post-pandemic group (n = 5648) and the pandemic group (n = 1391). Sociodemographic characteristics of the pandemic (CoronabaBY) and post-pandemic (JuFaBY) samples were analyzed for potential differences using Chi-square or Welch tests. To examine potential differences in the presence of problems ('problematic' vs. 'not problematic' according to the cut-off) in terms of crying and sleeping (crying/whining/sleeping subscale, SFS), feeding (feeding subscale, SFS), as well as excessive crying according to the 'Rule of Threes' (excessive crying, SFS) between the pandemic and post-pandemic groups, Chi-square tests were conducted. In case the cell frequencies were below 5, the Fisher Exact Test was calculated. Additionally, to assess the strength of the association between the variables, the effect size Phi (φ) was evaluated.

For the second research question, mothers and fathers in the post-pandemic sample were analyzed for possible differences in their demographic characteristics using the Chi-square or Welch test. We conducted three mediation models for each father and mother to investigate the relationship between overall infant regulatory problems (Infant Overall Regulatory Problems, SFS) and paternal and maternal parenting behavior (model 1: sensitivity subscale, CEQPAQ; model 2: responsiveness subscale, CEQPAQ; and model 3: over-reactivity subscale, CEQPAQ) and whether this relationship was mediated by paternal or maternal parenting stress (EBI Total Score T-values). The mediation analyses were performed using Hayes' free PROCESS tool. Potential heteroscedasticity in the data was controlled for using heteroscedasticity-consistent standard errors (HC3). This method accounts for the possible non-constancy of the error variance, improving the accuracy of standard error estimates. The mediation analysis included the calculation of the direct effects for the individual paths and the indirect mediation effect at a 5% significance level. In cases of full mediation, the total effect loses significance. If the significance of the relationship between the independent and dependent variables remains after the inclusion of the mediator but is reduced, partial mediation can be assumed. The mediation effect was evaluated using the bootstrap method with 5000 resampling iterations, and confidence intervals for the indirect effect were calculated.

Finally, a post hoc power analysis using G*Power (https://www.psychologie.hhu.de/arbeitsgruppen/allgemeine-psychologie-und-arbeitspsychologie/gpower (accessed on 4 September 2024)) was conducted to determine whether the power $1-\beta$ was within an acceptable range for small, medium, and large effects.

3. Results

3.1. Descriptives

The comparison between the samples showed that the parents in the post-pandemic (JuFaBY) group were, on average, younger than those in the pandemic group (CoronabaBY). Additionally, the post-pandemic group had a lower percentage of highly educated parents. The infant age was also, on average, lower in the post-pandemic group compared to that in the pandemic group. Due to differences in how nationality was recorded, the groups could not be compared on this variable. The effect sizes were consistently very small to small (see Table 1).

Table 1. Demographic characteristics of the post-pandemic group and pandemic group.

	Post-Pandemic Group (JuFaBY) (n = 5648)		Pandemic Group (CoronabaBY) (n = 1391)		Statistical Significance (Effect Size)
Parents					
Parental age (years), mean (*SD*)	32.06 (4.55)		32.75 (4.57)		[b] <0.001 * (0.15)
	%	n	%	n	

Table 1. *Cont.*

	Post-Pandemic Group (JuFaBY) (n = 5648)		Pandemic Group (CoronabaBY) (n = 1391)		Statistical Significance (Effect Size)
Fathers	5.6	314	6.8	95	[a] 0.053 (−0.02)
(At least one) parent born in Germany	96.0	5423	91.0	1266	n.a.
Single parent	7.4	420	7.5	105	[a] 0.864 (0.00)
Level of education					[a] <0.001 * (0.06)
High	59.5	3362	62.4	847	
Low	38.6	2178	37.6	511	
Others	1.9	108	2.4	33	
Children					
Mean infant age in months, (*SD*)		4.86 (3.83)		5.47 (3.51)	[b] <0.001 * (0.16)
	%	n	%	n	
Male	53.3	3013	51.8	720	[a] 0.297 (0.02)
Siblings	30.5	1722	43.6	607	[a] <0.001 * (−0.11)

[a] Chi-square test (effect size Phi ϕ); [b] Welch test (effect size Cohen's d); * $p \leq 0.05$.

Comparing the mothers and fathers of the post-pandemic group for the second research question, we found that the fathers were, on average, older than the mothers. Additionally, a slightly higher proportion of mothers reported that one or both parents were born in Germany compared to fathers. The level of education also differed significantly, with a higher percentage of fathers having a higher education. There was also a notable difference in the single parent status, with mothers being more likely to be single parents. In terms of the children, the average age of the infants was lower among fathers than mothers, and a higher proportion of mothers reported having other children (siblings of the infant) (see Table 2). The effect sizes were very small to small except for the age difference in the parents (medium effect).

Table 2. Demographic characteristics of mothers and fathers in the post-pandemic sample.

	Mothers (n = 5334)		Fathers (n = 314)		Statistical Significance (Effect Size)
Parents					
Parental age (years), mean (*SD*)		31.91 (4.47)		34.50 (5.24)	[b] <0.001 ** (−0.57)
	%	n	%	n	
Parent born in Germany	96.2	5129	93.6	294	[a] <0.001 ** (0.10)
Predominantly single parent	7.8	417	1.0	3	[a] <0.001 ** (−0.06)
Level of education					[a] <0.001 ** (0.08)
High	58.6	3126	75.2	236	
Low	39.4	2103	23.9	75	
Others	2.0	105	1.0	3	

Table 2. *Cont.*

	Mothers (n = 5334)		Fathers (n = 314)		Statistical Significance (Effect Size)
Children					
Infant age (months), mean (*SD*)	4.93 (3.85)		3.65 (3.37)		[b] <0.001 ** (0.34)
	%	n	%	n	
Male	53.4	2851	51.6	162	[a] 0.310 (0.02)
Siblings	31.2	1662	19.1	60	[a] <0.001** (−0.06)

[a] Chi-square test/Fisher Exact Test (effect size Phi ϕ); [b] Welch test (effect size Cohen's d); ** $p \leq 0.001$.

3.2. *Infant Crying, Sleeping, and Feeding Problems During and After the Pandemic*

According to the SFS, significantly more infants in the post-pandemic group showed problems in the areas of crying/whining/sleeping (χ^2 (1, n = 7039) = 7.83, p = 0.005). This was also the case for excessive crying defined by the 'Rule of Threes' (χ^2 (1, n = 7039) = 5.38, p = 0.020). The effect sizes were small. There were no significant differences between the two groups with regard to feeding problems. For detailed information, see Table 3.

Table 3. Differences in infant crying, sleeping, and feeding problems in the post-pandemic and pandemic samples.

Infant Mental Health (CSF) (Above Cut-Off)	Post-Pandemic Group (JuFaBY) (n = 5648)		Pandemic Group (CoronabaBY) (n = 1391)		Statistical Significance (Effect Size ϕ)
Noticeable crying, feeding, and sleeping problems	%	n	%	n	
Excessive crying (Wessel criterion)	6.3	357	4.7	65	0.020 * (0.03)
Crying/whining/sleeping	34.8	1966	30.8	429	0.005 * (0.03)
Feeding	35.1	1981	36.7	510	0.273 (−0.01)

Chi-square test; * $p \leq 0.05$.

3.3. *Parenting Stress and Parenting Behavior in Mothers and Fathers in the Post-Pandemic Sample*

The mean paternal parenting stress score in the parent domain of the EBI for this sample was 72.26 (*SD* = 19.30), and the mean maternal parenting stress score was 77.75 (*SD* = 20.34). The mean T-score was 55.42 (*SD* = 9.66) in fathers and 57.95 (*SD* = 9.61) in mothers, which, on average, fall within the ´not stressed´ range (<59). Additionally, the categorical classification based on the global stress score showed that 37.5% of fathers and 48.5% of mothers felt 'stressed' or 'strongly stressed' (see Table 4).

Table 4. Parenting stress in mothers and fathers in the post-pandemic sample.

Parenting Stress (EBI)	%	n	%	n
Categorial Evaluation of the Parent Domain	Fathers (n = 314)		Mothers (n = 5334)	
No findings	62.4	196	51.5	2749
Stressed	29.9	94	34.5	1840
Strongly stressed	7.6	24	14.0	745

Fathers scored an average of M = 24.18 (*SD* = 3.80) (mothers: M = 26.17, *SD* = 3.18) on the responsiveness subscale and M = 18.58 (*SD* = 2.94) on the sensitivity subscale (mothers: M = 20.07, *SD* = 2.71). On the over-reactivity subscale, the average paternal score was M = 8.70 (*SD* = 3.36) (mothers: M = 9.75, *SD* = 3.77).

3.4. Relationships Between Infant Crying, Sleeping, and Feeding Problems, Parenting Stress, and Parenting Behaviors in the Post-Pandemic Sample

3.4.1. Infant Crying, Sleeping, and Feeding Problems, Parenting Stress, and Sensitivity in Mothers and Fathers

After parenting stress was included as a mediator in the model, infant crying, sleeping, and feeding problems were found to significantly negatively predict sensitive parenting behavior (mothers: $\beta = -0.36$, $p < 0.001$; fathers: $\beta = -0.28$, $p < 0.001$) and positively predict parenting stress (mothers: $\beta = 0.47$, $p < 0.001$; fathers: $\beta = 0.43$, $p < 0.001$). Moreover, parenting stress was significantly negatively associated with sensitivity in both parents (mothers: $\beta = -0.23$, $p < 0.001$; fathers: $\beta = -0.32$, $p < 0.001$). The indirect effects of both models (mothers: $\beta = -0.11$, 95% CI [−1.225, −0.946]; fathers: $\beta = -0.14$, 95% CI [−0.2005, −0.0864]) indicate a significant mediation [56]. Despite the mediated effect, the direct effect of infant crying, sleeping, and feeding problems on sensitivity remained significant (mothers: $\beta = -0.25$, $p < 0.001$; fathers: $\beta = -0.15$, $p < 0.001$), suggesting a partial mediation (see Figures 1 and 2).

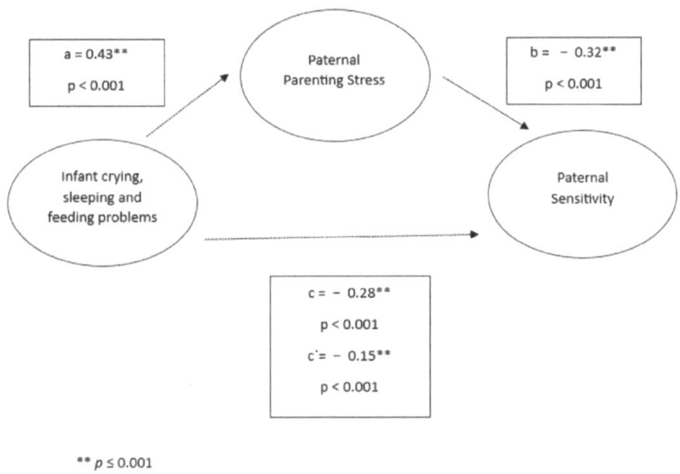

Figure 1. Infant crying, sleeping, and feeding problems, paternal parenting stress, and paternal sensitivity.

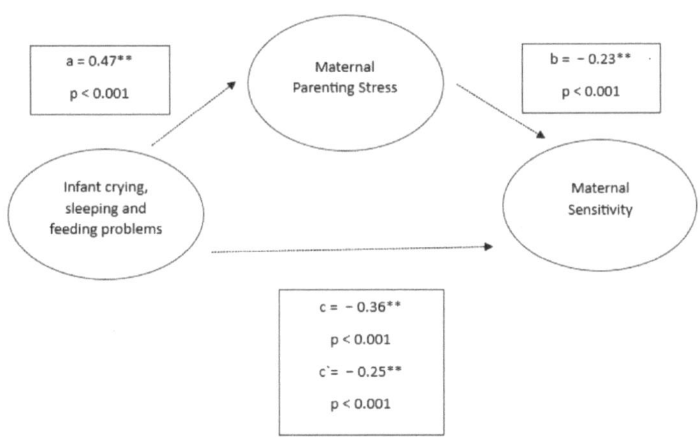

Figure 2. Infant crying, sleeping, and feeding problems, maternal parenting stress, and maternal sensitivity.

3.4.2. Infant Crying, Sleeping, and Feeding Problems, Parenting Stress, and Responsiveness in Mothers and Fathers

In the mediation models, infant crying, sleeping, and feeding problems were significantly associated with parenting stress and responsive parenting behavior in both mothers and fathers. Specifically, infant crying, sleeping, and feeding problems were positively associated with parenting stress (mothers: $\beta = 0.47$, $p < 0.001$; fathers: $\beta = 0.43$, $p < 0.001$) and negatively associated with responsive parenting behavior (mothers: $\beta = -0.38$, $p < 0.001$; fathers: $\beta = -0.30$, $p < 0.001$). Parenting stress was also negatively associated with responsiveness in both parents (mothers: $\beta = -0.26$, $p < 0.001$; fathers: $\beta = -0.32$, $p < 0.001$). Significant indirect effects were observed in both models, indicating a significant mediation (mothers: $\beta = -0.12$, 95% CI [−0.1353, −0.1076]; fathers: $\beta = -0.14$, 95% CI [−0.2068, −0.0837]). Despite the mediated effect, the direct effect of infant crying, sleeping, and feeding problems on responsiveness remained significant for both mothers ($\beta = -0.26$, $p < 0.001$) and fathers ($\beta = -0.16$, $p = 0.004$), indicating partial mediation (see Figures 3 and 4).

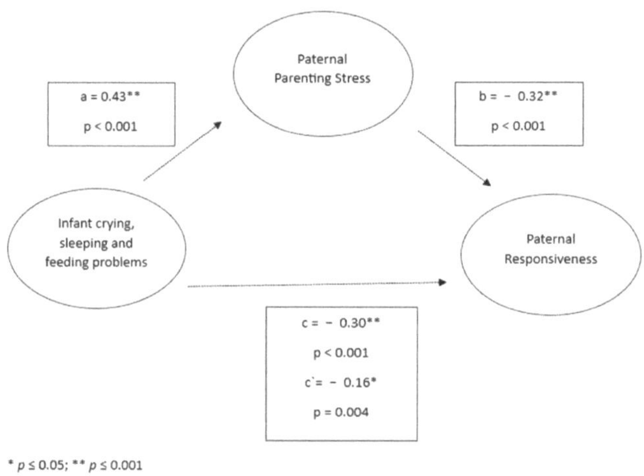

Figure 3. Infant crying, sleeping, and feeding problems, paternal parenting stress, and paternal responsiveness.

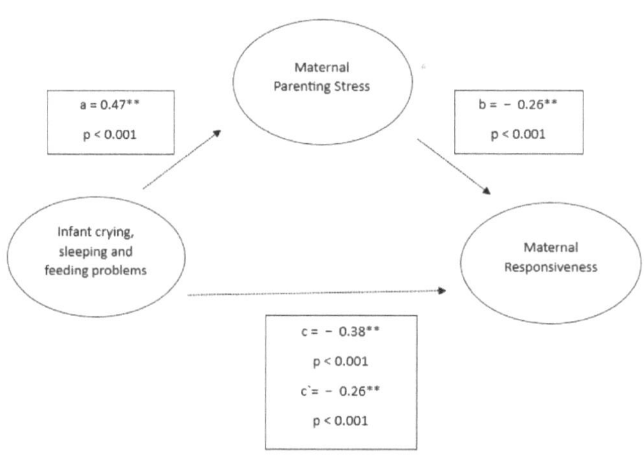

Figure 4. Infant crying, sleeping, and feeding problems, maternal parenting stress, and maternal responsiveness.

3.4.3. Infant Crying, Sleeping, and Feeding Problems, Parenting Stress, and Over-Reactivity in Mothers and Fathers

In the mediation model examining the effects of infant crying, sleeping, and feeding problems on parenting behaviors, parenting stress emerged as a significant mediator in the relationship between infant difficulties and over-reactive parenting for both mothers and fathers. Infant crying, sleeping, and feeding problems were positively associated with parenting stress (mothers: $\beta = 0.47$, $p < 0.001$; fathers: $\beta = 0.43$, $p < 0.001$), which, in turn, was associated with higher levels of over-reactive parenting behavior (mothers: $\beta = 0.42$, $p < 0.001$; fathers: $\beta = 0.41$, $p < 0.001$). Additionally, infant symptoms had a direct effect on over-reactivity, with significant values observed in both parents (mothers: $\beta = 0.31$, $p < 0.001$; fathers: $\beta = 0.34$, $p < 0.001$).

The indirect effect of infant crying, sleeping, and feeding problems on over-reactive parenting behavior through parenting stress was significant for both mothers and fathers (mothers: $\beta = 0.20$, 95% CI [0.1847, 0.2150]; fathers: $\beta = 0.18$, 95% CI [0.1176, 0.2465]). Despite this mediated effect, the direct effect of infant crying, sleeping, and feeding problems on over-reactivity remained significant for both parents (mothers: $\beta = 0.11$, $p < 0.001$; fathers: $\beta = 0.16$, $p = 0.003$), suggesting that parenting stress partially mediated the relationship between infant difficulties and over-reactive parenting across both the maternal and paternal models (see Figures 5 and 6).

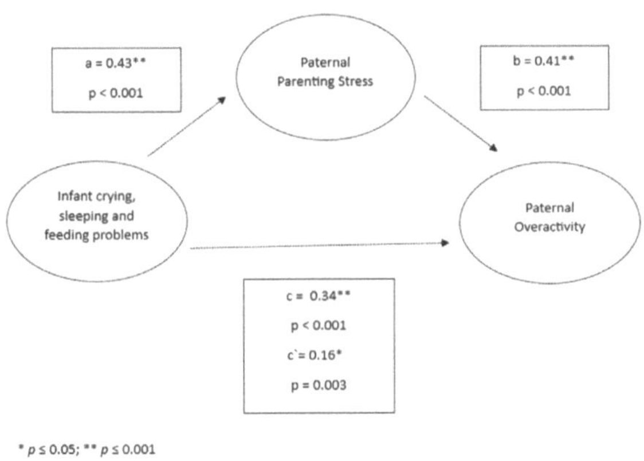

Figure 5. Infant crying, sleeping, and feeding problems, paternal parenting stress, and paternal over-reactivity.

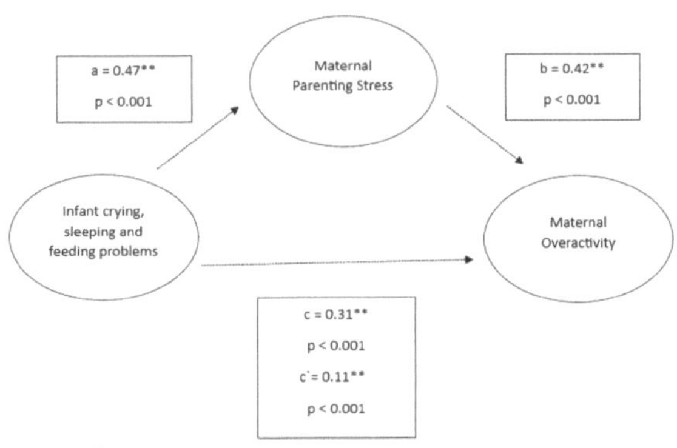

Figure 6. Infant crying, sleeping, and feeding problems, maternal parenting stress, and maternal over-reactivity.

4. Discussion

In a cross-sectional comparison study of pandemic (n = 1391) and post-pandemic (n = 5648) samples, including mothers and fathers with children aged 0–2 years, we examined whether the prevalence of infant crying, sleeping, and feeding problems differed between these two periods and whether parenting stress mediated the relationship between infant mental health problems and parenting behavior in the post-pandemic group. Our findings indicated significant differences in the prevalence of infant crying and sleeping problems as well as excessive crying between the pandemic and post-pandemic samples, suggesting that infant regulatory problems remain notably pronounced and are partially even more highly evident than during the pandemic. Maternal and paternal parenting stress partially mediated the association between infant crying, sleeping, and feeding problems and parenting behaviors with consistently small observed effects.

Examining our results in detail, the comparison with the pandemic reference sample revealed that significantly more crying/whining/sleeping problems and excessive crying were reported for infants of the post-pandemic group. Infant crying/whining/sleeping problems and excessive crying were reported by 34.8% and 6.3% of the parents in the post-pandemic sample, respectively. Feeding difficulties appeared in 35.1% of the infants in this group. Since regulatory problems are known to be transient in infancy and early childhood and vary significantly according to the age and developmental stage, classifications remain challenging. Yet, we found the rates to be relatively high in comparison to pre-pandemic reference studies in non-clinical samples, where the incidences for all these problems usually range up to around 20% [12,57–61]. The observed proportion of infants with regulatory problems also exceeded the levels found during the pandemic in prior studies, suggesting that infant mental health is particularly vulnerable in times of societal crises.

Theoretical explanations for our result include the fact that infants are particularly susceptible to environmental influences, with primary caregivers and the family environment serving as dominant influencing factors in their early development [62,63]. Current studies on parenting stress indicate that the parenting burden remains alarmingly high post-pandemic [42]. The prolonged period of anxiety and uncertainty, e.g., due to fear of infection, disrupted daily routines, and childcare closures, during the pandemic might have led to a state of mental exhaustion in parents [64]. Furthermore, the long-term effects of the crisis—despite the pandemic being considered over—may be compounding with the impacts of new societal challenges, such as the existential threat of climate change, economic inflation, and the wars in Ukraine and the Middle East. Today's families are placed under unprecedented strain, which might lead to potential fatigue due to a cumulation of social challenges. Societal crises also represent a risk factor for parental anxiety and depression [42], which, in turn, are known to have potential negative effects on children's mental health [65]. However, the influence of the present societal crises on infant mental health will have to be investigated in future studies since we cannot confirm these ideas on the basis of our data.

Our mediation models revealed that infant crying, sleeping, and feeding problems were linked to reduced maternal and paternal sensitivity and responsiveness along with increased over-reactive behaviors in both parents. Due to the lack of comparable studies with corresponding models, the individual paths can primarily be classified within the existing literature. Research has demonstrated that positive parenting behaviors can support and strengthen children's regulatory capacities, suggesting that sensitive and responsive parental actions positively impact child regulation [66]. Conversely, evidence also indicates a bidirectional relationship, whereby child and parent behaviors mutually shape one another. For instance, regulatory problems in children may contribute to increased parental uncertainty, especially when conventional soothing methods prove ineffective, potentially diminishing parental confidence. This interplay can, in turn, influence parental behavior, which may inadvertently reinforce regulatory problems in the infant [67,68]. Since the design of this study does not allow for conclusions about the possible causal directions, further longitudinal studies should be conducted in this regard.

Additionally, infant crying, sleeping, and feeding difficulties positively predicted parenting stress in our study. In line with studies on parenting stress related to regulatory problems [38,69], a heightened infant symptomatology was associated with increased parenting stress. Furthermore, stressed parents demonstrated lower sensitivity and responsiveness, paired with more over-reactive parenting behaviors. These findings also add to the existing literature on parenting stress and behavior in parents, which has shown that higher stress levels correlate with harsher and more controlling parenting practices [40,70,71]. Parenting stress can also impair the ability to perceive and respond accurately to children's needs [33,72]. In our study, maternal and paternal parenting stress partially mediated the link between infant crying, sleeping, and feeding problems and parenting behaviors. The models for mothers and fathers aligned in the direction of direct paths and demonstrated comparable effect sizes, with consistently small mediation effects in both cases, suggesting that these relationships may be similarly applicable for both parents. This implies that infant crying, sleeping, and feeding difficulties could partly contribute to increased parenting stress for both mothers and fathers, potentially reducing their parenting capacity by lowering their sensitivity and responsiveness while increasing their over-reactivity. While fathers still play a minor role in the research related to infant mental health, during the pandemic, noticeable changes emerged in family dynamics, including an increased involvement of fathers in the daily care and upbringing of their children. Fathers tend to spend more time in the everyday life of their children [45]. This makes it all the more important to bring fathers and their influence on the family system into greater focus in future research.

The interplay between infant regulatory problems, parenting stress, and parenting behavior is central to understanding the dynamics surrounding crying, sleeping, and feeding difficulties within the family system. The reciprocal nature of these challenges indicates that while infant difficulties can impact parental sensitivity and responsiveness, parenting behaviors, in turn, influence the course of these issues [66]. Recognizing this cycle provides valuable opportunities for intervention, potentially breaking the loop and lessening the impact of regulatory problems on both parents and infants, thereby increasing the chances of symptom relief. Given that parents often seek support only once regulatory issues are already present, we consider it particularly relevant in clinical practice to view already existing infant regulatory problems as a starting point for exploring the complex mechanisms within the family system.

Some strengths and limitations need to be considered when interpreting the underlying results. While the pandemic and post-pandemic samples were relatively similar in structure, they differed in terms of the parental age, infant age, presence of siblings, and educational background. We decided against controlling for these factors due to the very small to small effect sizes and also based on theoretical considerations: to the best of our knowledge, no studies specifically link parental age to infant regulatory problems. Although findings regarding the influence of having siblings/the birth order are somewhat inconsistent, most of the available data suggest no significant effects on infant crying or sleep patterns [73–75]. Evidence on the association between parental education and infant regulatory issues is very limited; to the best of our knowledge, there is one study that indicates that a lower educational background may be a risk factor for infant crying, sleeping, and feeding problems [76]. However, this study was from Turkey and this result cannot be fully transferred to the German context. In addition, the total sample consisted of mainly high educated parents, making it less likely to act as a confounding factor.

Differences also emerged with respect to the child's age. Typically, the crying duration increases in the initial postnatal weeks, peaking around four to six weeks and then declining at around three to four months [77,78]. Since both samples, on average, were beyond this age range, age-related differences in crying behavior were not expected. Infant sleep patterns are variable and continually developing, but they generally start to stabilize at around five to six months, with longer nighttime sleep and increased sleep consolidation [79]. Both groups fell within this age range, reducing the likelihood of age-related effects.

The sample of mothers and fathers in the post-pandemic sample also differed significantly in terms of single parenthood, their educational level, and the presence of additional children. Yet, the inherent differences between mothers and fathers, especially with regard to being a single parent and education, are a reality that must be faithfully represented in the research. Accurately capturing these distinctions is crucial for producing findings that genuinely reflect the complexities of real-life dynamics. Overall, the sample primarily consisted of highly educated parents from a German background, resulting in a limited generalizability of the findings to other populations. It should also be considered that the mothers and fathers in the sample did not come from the same families; therefore, it remains unclear whether the perception of the child symptoms of one parent is influenced by the other parent.

Furthermore, the analysis is based on a cross-sectional design, which limits the ability to draw causal conclusions regarding the relationships between infant regulatory problems, parenting stress, and parenting behavior. Additionally, the small effect sizes observed in the mediation models suggest that the findings should not be overinterpreted. While they provide important initial insights, they do not yet establish a clear, clinically relevant effect. Nevertheless, this study represented an initial investigation of constructs that increase the understanding of infant regulatory problems as a triad of the child behavior, parenting burden, and parenting behavior within a single model and for both mothers and fathers of infants. As such, it provides valuable starting points and insights for further research in this area. The use of self-report measures for assessing infant symptoms and parenting behavior is another limitation, though standardized and validated instruments were employed.

Despite these limitations, the study presents several strengths. First, our study offers a comparison between distinct pandemic and post-pandemic groups with a considerably high sample size, providing valuable insights into infant mental health during and after the pandemic. Furthermore, this study is one of the first to examine the role of both mothers and fathers in the context of infant regulatory problems, particularly in light of the changes in family structures during the pandemic.

5. Conclusions

In conclusion, our findings underscore the continued significance of addressing infant mental health in the post-pandemic period. In light of the potentially lasting effects of the pandemic on parents and, subsequently, their young children, and the increasing influence of various societal challenges, it is imperative that both research and practice place greater emphasis on the wellbeing of infants and their families. Additionally, our results provide valuable insights into the complex interplay between infant regulatory problems, parenting stress, and parenting behaviors within the broader family system. A comprehensive understanding of these inter-related factors is, amongst others, crucial for developing effective support strategies aimed at enhancing the wellbeing of both the infant and the family as a whole. In particular, parenting stress—stemming from infant behavioral problems—emerges as a potential entry point for addressing the cycle of infant regulatory problems and maladaptive behaviors in both fathers and mothers.

Author Contributions: Conceptualization, A.F. and V.M.; data curation, K.R.; formal analysis, K.R.; funding acquisition, V.M. and A.F.; investigation, A.F. and M.A.; methodology, A.F., K.R. and M.A.; project administration, A.F. and V.M.; resources, V.M.; supervision, M.A.; validation, A.F. and M.A.; writing—original draft, A.F. and K.R.; writing—review and editing, V.M. and M.A. All authors have read and agreed to the published version of the manuscript.

Funding: The authors declare that financial support was received for the research, authorship, and/or publication of this article. This study was funded by the Bavarian State Ministry of Family, Labour and Social Affairs (Bayerisches Staatsministerium für Familie, Arbeit und Soziales) and by the Federal Ministry of Education and Research (Bundesministerium für Bildung und Forschung, BMBF) as part of the German Center for Child and Adolescent Health (DZKJ) under the funding code 01GL2406B.

Institutional Review Board Statement: The study presented in our manuscript is based on datasets from two projects, and the respective approvals by the Ethics Committee of the Technical University of Munich are as follows: CoronabaBY: 322/20 S (date of approval: 24 September 2020) and JuFaBY: 2022-483-S-KH (date of approval: 13 September 2022).

Informed Consent Statement: Informed consent was obtained from all subjects involved in this study.

Data Availability Statement: The data necessary to reproduce the analyses presented here are not publicly accessible.

Acknowledgments: First, we thank all the pediatricians and families who participated in the CoronabaBY and JuFaBY studies despite these challenging and particularly stressful times. This study could not have been realized without the support of BVKJ Bayern (Professional Association of Pediatricians in Bavaria) and Paednetz Bayern—represented by Dominik Ewald, Otto Laub, and Gabi Haus—as well as Monks Aerzte-im-Netz GmbH and BVKJ Service GmbH, who also made the implementation of this digital study possible. We also thank the CoronabaBY and JuFaBY research teams, particularly Ina Nehring and Catherine Buechel. Special thanks go to Maria Licata for her valuable clinical expertise on infant crying, sleeping, and feeding problems.

Conflicts of Interest: The authors declare that they do not have any conflicts of interest.

References

1. Egle, U.T.; Franz, M.; Joraschky, P.; Lampe, A.; Seiffge-Krenke, I.; Cierpka, M. Gesundheitliche Langzeitfolgen psychosozialer Belastungen in der Kindheit—Ein Update. *Bundesgesundheitsblatt Gesundheitsforschung Gesundheitsschutz* **2016**, *59*, 1247–1254. [CrossRef]
2. Laucht, M.; Schmidt, M.H.; Esser, G. Motorische, kognitive und sozial-emotionale Entwicklung von 11-Jährigen mit frühkindlichen Risikobelastungen: Späte Folgen. *Z. Kinder Jugendpsychiatr. Psychother.* **2002**, *30*, 5–19. [CrossRef]
3. Heilig, L. Risikokonstellationen in der frühen Kindheit: Auswirkungen biologischer und psychologischer Vulnerabilitäten sowie psychosozialer Stressoren auf kindliche Entwicklungsverläufe. *Z. Erzieh.* **2014**, *17*, 263–280. [CrossRef]
4. Huebener, M.; Waights, S.; Spiess, C.K.; Siegel, N.A.; Wagner, G.G. Parental well-being in times of COVID-19 in Germany. *Rev. Econ. Househ.* **2021**, *19*, 91–122. [CrossRef] [PubMed]
5. Panda, P.K.; Gupta, J.; Chowdhury, S.R.; Kumar, R.; Meena, A.K.; Madaan, P.; Sharawat, I.K.; Gulati, S. Psychological and Behavioral Impact of Lockdown and Quarantine Measures for COVID-19 Pandemic on Children, Adolescents and Caregivers: A Systematic Review and Meta-Analysis. *J. Trop. Pediatr.* **2021**, *67*, fmaa122. [CrossRef]
6. Chawla, N.; Tom, A.; Sen, M.S.; Sagar, R. Psychological Impact of COVID-19 on Children and Ad-olescents: A Systematic Review. *Indian J. Psychol. Med.* **2021**, *43*, 294–299. [CrossRef] [PubMed]
7. Schlack, R.; Neuperdt, L.; Junker, S.; Eicher, S.; Hölling, H.; Thom, J.; Ravens-Sieberer, U.; Beyer, A.-K. Changes in mental health in the German child and adolescent population during the COVID-19 pandemic—Results of a rapid review. *J. Health Monit.* **2023**, *8* (Suppl. 1), 2–72.
8. Lopatina, O.L.; Panina, Y.A.; Malinovskaya, N.A.; Salmina, A.B. Early life stress and brain plasticity: From molecular alterations to aberrant memory and behavior. *Rev. Neurosci.* **2021**, *32*, 131–142. [CrossRef] [PubMed]
9. Brown, T.T.; Jernigan, T.L. Brain development during the preschool years. *Neuropsychol. Rev.* **2012**, *22*, 313–333. [CrossRef]
10. Knickmeyer, R.C.; Gouttard, S.; Kang, C.; Evans, D.; Wilber, K.; Smith, J.K.; Hamer, R.M.; Lin, W.; Gerig, G.; Gilmore, J.H. A structural MRI study of human brain development from birth to 2 years. *J. Neurosci.* **2008**, *28*, 12176–12182. [CrossRef] [PubMed]
11. Ziegler, M.; Wollwerth de Chuquisengo, R.; Mall, V.; Licata-Dandel, M. Frühkindliche psychische Störungen: Exzessives Schreien, Schlaf- und Fütterstörungen sowie Interventionen am Beispiel des "Münchner Modells". *Bundesgesundheitsblatt-Gesundheitsforschung-Gesundheitsschutz* **2023**, *66*, 752–760. [CrossRef] [PubMed]
12. Bilgin, A.; Wolke, D. Regulatory Problems in Very Preterm and Full-Term Infants Over the First 18 Months. *J. Dev. Behav. Pediatr.* **2016**, *37*, 298–305. [CrossRef]
13. Cooke, J.E.; Kochendorfer, L.B.; Stuart-Parrigon, K.L.; Koehn, A.J.; Kerns, K.A. Parent-child attachment and children's experience and regulation of emotion: A meta-analytic review. *Emotion* **2019**, *19*, 1103–1126. [CrossRef]
14. Petzoldt, J.; Wittchen, H.-U.; Einsle, F.; Martini, J. Maternal anxiety versus depressive disorders: Specific relations to infants' crying, feeding and sleeping problems. *Child Care Health Dev.* **2016**, *42*, 231–245. [CrossRef] [PubMed]
15. Rask, C.U.; Ørnbøl, E.; Olsen, E.M.; Fink, P.; Skovgaard, A.M. Infant behaviors are predictive of functional somatic symptoms at ages 5–7 years: Results from the Copenhagen Child Cohort CCC2000. *J. Pediatr.* **2013**, *162*, 335–342. [CrossRef]
16. Sidor, A.; Fischer, C.; Cierpka, M. The link between infant regulatory problems, temperament traits, maternal depressive symptoms and children's psychopathological symptoms at age three: A lon-gitudinal study in a German at-risk sample. *Child Adolesc. Psychiatry Ment. Health* **2017**, *11*, 10. [CrossRef] [PubMed]
17. Galling, B.; Brauer, H.; Struck, P.; Krogmann, A.; Gross-Hemmi, M.; Prehn-Kristensen, A.; Mudra, S. The impact of crying, sleeping, and eating problems in infants on childhood behavioral outcomes: A meta-analysis. *Front. Child Adolesc. Psychiatry* **2023**, *1*, 1099406. [CrossRef]

18. Perez, A.; Göbel, A.; Stuhrmann, L.Y.; Schepanski, S.; Singer, D.; Bindt, C.; Mudra, S. Born Under COVID-19 Pandemic Conditions: Infant Regulatory Problems and Maternal Mental Health at 7 Months Postpartum. *Front. Psychol.* **2021**, *12*, 805543. [CrossRef] [PubMed]
19. Sprengeler, M.K.; Mattheß, J.; Galeris, M.-G.; Eckert, M.; Koch, G.; Reinhold, T.; Berghöfer, A.; Fricke, J.; Roll, S.; Keil, T.; et al. Being an Infant in a Pandemic: Influences of the COVID-19 Pandemic on Infants, Toddlers and Their Mothers in a Clinical Population. *Children* **2023**, *10*, 1885. [CrossRef] [PubMed]
20. Bradley, H.; Fine, D.; Minai, Y.; Gilabert, L.; Gregory, K.; Smith, L.; Gao, W.; Giase, G.; Krogh-Jespersen, S.; Zhang, Y.; et al. Maternal perceived stress and infant behavior during the COVID-19 pandemic. *Pediatr. Res.* **2023**, *94*, 2098–2104. [CrossRef] [PubMed]
21. Nazzari, S.; Pili, M.P.; Günay, Y.; Provenzi, L. Pandemic babies: A systematic review of the asso-ciation between maternal pandemic-related stress during pregnancy and infant development. *Neurosci. Biobehav. Rev.* **2024**, *162*, 105723. [CrossRef] [PubMed]
22. Provenzi, L.; Grumi, S.; Altieri, L.; Bensi, G.; Bertazzoli, E.; Biasucci, G.; Cavallini, A.; Decembrino, L.; Falcone, R.; Freddi, A.; et al. Prenatal maternal stress during the COVID-19 pandemic and infant regulatory capacity at 3 months: A longitudinal study. *Dev. Psychopathol.* **2023**, *35*, 35–43. [CrossRef]
23. Wiley, K.S.; Fox, M.M.; Gildner, T.E.; Thayer, Z.M. A longitudinal study of how women's prenatal and postnatal concerns related to the COVID-19 pandemic predicts their infants' social-emotional development. *Child Dev.* **2023**, *94*, 1356–1367. [CrossRef] [PubMed]
24. Reinelt, T.; Suppiger, D.; Frey, C.; Oertel, R.; Natalucci, G. Infant regulation during the pandemic: Associations with maternal response to the COVID-19 pandemic, well-being, and socio-emotional investment. *Infancy* **2023**, *28*, 9–33. [CrossRef]
25. Buechel, C.; Nehring, I.; Seifert, C.; Eber, S.; Behrends, U.; Mall, V.; Friedmann, A. A cross-sectional investigation of psychosocial stress factors in German families with children aged 0-3 years during the COVID-19 pandemic: Initial results of the CoronabaBY study. *Child Adolesc. Psychiatry Ment. Health* **2022**, *16*, 37. [CrossRef]
26. Louie, P.; Upenieks, L.; Hill, T.D. Cumulative Pandemic Stressors, Psychosocial Resources, and Psychological Distress: Toward a More Comprehensive Test of a Pandemic Stress Process. *Soc. Ment. Health* **2023**, *13*, 245–260. [CrossRef]
27. Clayton, S. Climate anxiety: Psychological responses to climate change. *J. Anxiety Disord.* **2020**, *74*, 102263. [CrossRef]
28. Lass-Hennemann, J.; Sopp, M.R.; Ruf, N.; Equit, M.; Schäfer, S.K.; Wirth, B.E.; Michael, T. Generation climate crisis, COVID-19, and Russia-Ukraine-War: Global crises and mental health in adolescents. *Eur. Child Adolesc. Psychiatry* **2024**, *33*, 2203–2216. [CrossRef]
29. Movsisyan, A.; Wendel, F.; Bethel, A.; Coenen, M.; Krajewska, J.; Littlecott, H.; Stöckl, H.; Voss, S.; Wollmershäuser, T.; Rehfuess, E. Inflation and health: A global scoping review. *Lancet Glob. Health* **2024**, *12*, e1038–e1048. [CrossRef]
30. Malia, J.A. Basic Concepts and Models of Family Stress. *Stress Trauma Crisis* **2006**, *9*, 141–160. [CrossRef]
31. Georg, A.K.; Bark, C.; Wiehmann, J.; Taubner, S. Frühkindliche Regulationsstörungen: Störungsbilder und Behandlungskonzepte. *Psychotherapeut* **2022**, *67*, 265–278. [CrossRef]
32. Cooke, J.E.; Deneault, A.-A.; Devereux, C.; Eirich, R.; Fearon, R.M.P.; Madigan, S. Parental sensitivity and child behavioral problems: A meta-analytic review. *Child Dev.* **2022**, *93*, 1231–1248. [CrossRef]
33. Ward, K.P.; Lee, S.J. Mothers' and Fathers' Parenting Stress, Responsiveness, and Child Wellbeing Among Low-Income Families. *Child. Youth Serv. Rev.* **2020**, *116*, 105218. [CrossRef]
34. Deans, C.L. Maternal sensitivity, its relationship with child outcomes, and interventions that address it: A sys-tematic literature review. *Early Child Dev. Care* **2018**, *190*, 252–275. [CrossRef]
35. Goldstein, L.H.; Harvey, E.A.; Friedman-Weieneth, J.L. Examining subtypes of behavior problems among 3-year-old children, Part III: Investigating differences in parenting practices and parenting stress. *J. Abnorm. Child Psychol.* **2007**, *35*, 125–136. [CrossRef]
36. Abidin, R.R. Introduction to the Special issue: The Stresses of Parenting. *J. Clin. Child Psychol.* **1990**, *19*, 298–301. [CrossRef]
37. Beebe, S.A.; Casey, R.; Pinto-Martin, J. Association of reported infant crying and maternal parenting stress. *Clin. Pediatr.* **1993**, *32*, 15–19. [CrossRef] [PubMed]
38. Postert, C.; Averbeck-Holocher, M.; Achtergarde, S.; Müller, J.M.; Furniss, T. Regulatory disorders in early childhood: Correlates in child behavior, parent-child relationship, and parental mental health. *Infant Ment. Health J.* **2012**, *33*, 173–186. [CrossRef] [PubMed]
39. Chung, G.; Tilley, J.L.; Netto, N.; Chan, A.; Lanier, P. Parenting stress and its impact on parental and child functioning during the COVID-19 pandemic: A meta-analytical review. *Int. J. Stress Manag.* **2024**, *31*, 238–251. [CrossRef]
40. Chung, G.; Lanier, P.; Wong, P.Y.J. Mediating Effects of Parental Stress on Harsh Parenting and Parent-Child Relationship during Coronavirus (COVID-19) Pandemic in Singapore. *J. Fam. Violence* **2022**, *37*, 801–812. [CrossRef]
41. Friedmann, A.; Buechel, C.; Seifert, C.; Eber, S.; Mall, V.; Nehring, I. Easing pandemic-related re-strictions, easing psychosocial stress factors in families with infants and toddlers? Cross-sectional results of the three wave CoronabaBY study from Germany. *Child Adolesc. Psychiatry Ment. Health* **2023**, *17*, 76. [CrossRef] [PubMed]
42. Richter, K.; Buechel, C.; Augustin, M.; Friedmann, A.; Mall, V.; Nehring, I. Psychosocial stress in young families after the pandemic: No time to rest. *Res. Sq.* **2024**. [CrossRef]
43. Oberndorfer, R.; Rost, H. Neue Väter—Anspruch und Realität. *Z. Fam.* **2005**, *17*, 50–65.

44. Statistisches Bundesamt. Elterngeld 2022: Väteranteil Steigt Weiter auf 26.1%. 2022. Available online: https://www.destatis.de/DE/Presse/Pressemitteilungen/2023/03/PD23_123_22922.html (accessed on 6 November 2024).
45. Cito, G.; Micelli, E.; Cocci, A.; Polloni, G.; Coccia, M.E.; Carini, M.; Minervini, A.; Natali, A. Paternal Behaviors in the Era of COVID-19. *World J. Men's Health* **2020**, *38*, 251–253. [CrossRef]
46. Margaria, A. Fathers, Childcare and COVID-19. *Fem. Leg. Stud.* **2021**, *29*, 133–144. [CrossRef]
47. Weissbourd, R.; Batanova, M.; McIntyre, J.; Torres, E. How the Pandemic Is Strengthening Fathers' Relationships with Their Children. 2020. Available online: https://mcc.gse.harvard.edu/s/report-how-the-pandemic-is-strengthening-fathers-relationships-with-their-children-final.pdf (accessed on 10 November 2024).
48. Kellner, R. Die erschöpfte Mutter. *Z. Psychodrama Soziom.* **2024**, *23*, 315–327. [CrossRef]
49. Buechel, C.; Friedmann, A.; Eber, S.; Behrends, U.; Mall, V.; Nehring, I. The change of psychosocial stress factors in families with infants and toddlers during the COVID-19 pandemic. A longitudinal perspective on the CoronabaBY study from Germany. *Front. Pediatr.* **2024**, *12*, 1354089. [CrossRef]
50. Schmidtke, C.; Kuntz, B.; Starker, A.; Lampert, T. *Inanspruchnahme der Früherkennungsuntersuchungen für Kinder in Deutschland—Querschnittergebnisse aus KiGGS Welle 2*; Robert Koch-Institut: Berlin, Germany, 2018.
51. Gross, S.; Reck, C.; Thiel-Bonney, C.; Cierpka, M. Empirische Grundlagen des Fragebogens zum Schreien, Füttern und Schlafen (SFS). *Prax. Kinderpsychol. Kinderpsychiatr.* **2013**, *62*, 327–347. [CrossRef]
52. Wessel, M.A.; Cobb, J.C.; Jackson, E.B.; Harris, G.S.; Detwiler, A.C. Paroxysmal fussing in infancy, sometimes called "colic". *Pediatrics* **1954**, *14*, 421–435. [CrossRef]
53. Tröster, H. *Eltern-Belastungs-Inventar (EBI): Deutsche Version des Parenting Stress Index (PSI)*; Hogrefe: St. Newburyport, MA, USA, 2010.
54. Testzentrale. EBI. Eltern-Belastungs-Inventar. 2010. Available online: https://www.testzentrale.de/shop/eltern-belastungs-inventar.html (accessed on 13 October 2024).
55. Verhoeven, M.; Deković, M.; Bodden, D.; van Baar, A.L. Development and initial validation of the comprehensive early childhood parenting questionnaire (CECPAQ) for parents of 1–4 year-olds. *Eur. J. Dev. Psychol.* **2017**, *14*, 233–247. [CrossRef]
56. Preacher, K.J.; Hayes, A.F. SPSS and SAS procedures for estimating indirect effects in simple mediation models. *Behav. Res. Methods Instrum. Comput.* **2004**, *36*, 717–731. [CrossRef]
57. Lindberg, L.; Bohlin, G.; Hagekull, B. Early feeding problems in a normal population. *Int. J. Eat. Disord.* **1991**, *10*, 395–405. [CrossRef]
58. Lorenz, S.; Ulrich, S.M.; Sann, A.; Liel, C. Self-Reported Psychosocial Stress in Parents with Small Children. *Dtsch. Arztebl. Int.* **2020**, *117*, 709–716. [CrossRef] [PubMed]
59. Olsen, A.L.; Ammitzbøll, J.; Olsen, E.M.; Skovgaard, A.M. Problems of feeding, sleeping and excessive crying in infancy: A general population study. *Arch. Dis. Child.* **2019**, *104*, 1034–1041. [CrossRef]
60. Wake, M.; Morton-Allen, E.; Poulakis, Z.; Hiscock, H.; Gallagher, S.; Oberklaid, F. Prevalence, stability, and outcomes of cry-fuss and sleep problems in the first 2 years of life: Prospective communi-ty-based study. *Pediatrics* **2006**, *117*, 836–842. [CrossRef] [PubMed]
61. Wolke, D.; Schmid, G.; Schreier, A.; Meyer, R. Crying and feeding problems in infancy and cognitive outcome in preschool children born at risk: A prospective population study. *J. Dev. Behav. Pediatr.* **2009**, *30*, 226–238. [CrossRef] [PubMed]
62. Ainsworth, M.D.S.; Blehar, M.C.; Waters, E.; Wall, S.N. *Patterns of Attachment*; Psychology Press: London, UK, 2015.
63. Sroufe, L.A. Early relationships and the development of children. *Infant Ment. Health J.* **2000**, *21*, 67–74. [CrossRef]
64. Moradian, S.; Bäuerle, A.; Schweda, A.; Musche, V.; Kohler, H.; Fink, M.; Weismüller, B.; Benecke, A.V.; Dörrie, N.; Skoda, E.M.; et al. Differences and simi-larities between the impact of the first and the second COVID-19-lockdown on mental health and safety behaviour in Germany. *J. Public Health* **2021**, *43*, 710–713. [CrossRef]
65. Petzoldt, J. Systematic review on maternal depression versus anxiety in relation to excessive infant crying: It is all about the timing. *Arch. Women's Ment. Health* **2018**, *21*, 15–30. [CrossRef] [PubMed]
66. Samdan, G.; Kiel, N.; Petermann, F.; Rothenfußer, S.; Zierul, C.; Reinelt, T. The relationship between parental behavior and infant regulation: A systematic review. *Dev. Rev.* **2020**, *57*, 100923. [CrossRef]
67. Kim, J.; Cicchetti, D. Longitudinal pathways linking child maltreatment, emotion regulation, peer relations, and psychopathology. *J. Child Psychol. Psychiatry* **2010**, *51*, 706–716. [CrossRef]
68. Philbrook, L.E.; Teti, D.M. Bidirectional associations between bedtime parenting and infant sleep: Parenting quality, parenting practices, and their interaction. *J. Fam. Psychol.* **2016**, *30*, 431–441. [CrossRef] [PubMed]
69. Georg, A.K.; Schröder-Pfeifer, P.; Cierpka, M.; Taubner, S. Maternal Parenting Stress in the Face of Early Regulatory Disorders in Infancy: A Machine Learning Approach to Identify What Matters Most. *Front. Psychiatry* **2021**, *12*, 663285. [CrossRef] [PubMed]
70. Le, Y.; Fredman, S.J.; Feinberg, M.E. Parenting stress mediates the association between negative affectivity and harsh parenting: A longitudinal dyadic analysis. *J. Fam. Psychol.* **2017**, *31*, 679–688. [CrossRef] [PubMed]
71. Mak, M.C.K.; Yin, L.; Li, M.; Cheung, R.Y.; Oon, P.-T. The Relation between Parenting Stress and Child Behavior Problems: Negative Parenting Styles as Mediator. *J. Child Fam. Stud.* **2020**, *29*, 2993–3003. [CrossRef]
72. Deater-Deckard, K.; Panneton, R. *Parental Stress and Early Child Development*; Springer International Publishing: Cham, Switzerland, 2017.

73. Kaley, F.; Reid, V.; Flynn, E. Investigating the biographic, social and temperamental correlates of young infants' sleeping, crying and feeding routines. *Infant Behav. Dev.* **2012**, *35*, 596–605. [CrossRef]
74. Thunström, M. Severe sleep problems among infants in a normal population in Sweden: Prevalence, severity and correlates. *Acta Pædiatrica* **1999**, *88*, 1356–1363. [CrossRef]
75. Weissbluth, M. Naps in children: 6 months–7 years. *Sleep* **1995**, *18*, 82–87. [CrossRef] [PubMed]
76. Yalçin, S.S.; Orün, E.; Mutlu, B.; Madendağ, Y.; Sinici, I.; Dursun, A.; Özkara, H.A.; Üstünyurt, Z.; Kutluk, Ş.; Yurdakök, K. Why are they having infant colic? A nested case-control study. *Paediatr. Perinat. Epidemiol.* **2010**, *24*, 584–596. [CrossRef] [PubMed]
77. Brazelton, T.B. Crying in infancy. *Pediatrics* **1962**, *29*, 579–588. [CrossRef]
78. St James-Roberts, I.; Halil, T. Infant crying patterns in the first year: Normal community and clinical findings. *J. Child Psychol. Psychiatry* **1991**, *32*, 951–968. [CrossRef] [PubMed]
79. Mindell, J.A.; Leichman, E.S.; Composto, J.; Lee, C.; Bhullar, B.; Walters, R.M. Development of infant and toddler sleep patterns: Real-world data from a mobile application. *J. Sleep Res.* **2016**, *25*, 508–516. [CrossRef] [PubMed]

children MDPI

Article

The Experiences of Australian School Mental Health Professionals during COVID-19 Lockdowns

Emily Berger [1,2,*], Grace Mackie [1], Andrea Reupert [1], Daliya Greenfeld [1], Kelly-Ann Allen [1,3], Fiona May [1], Gerald Wurf [1], Dianne Summers [1] and Zoe Morris [1]

[1] School of Educational Psychology and Counselling, Faculty of Education, Monash University, Clayton, VIC 3800, Australia; grace@leveluppsychology.com.au (G.M.); andrea.reupert@monash.edu (A.R.); daliya.greenfeld@monash.edu (D.G.); kelly-ann.allen@monash.edu (K.-A.A.); fiona.may@monash.edu (F.M.); gerald.wurf@monash.edu (G.W.); dianne.summers@monash.edu (D.S.); zoe.morris@monash.edu (Z.M.)

[2] School of Rural Health, Faculty of Medicine, Nursing and Health Sciences, Monash University, Clayton, VIC 3800, Australia

[3] Centre for Wellbeing Science, Melbourne Graduate School of Education, University of Melbourne, Parkville, VIC 3010, Australia

* Correspondence: emily.berger@monash.edu; Tel.: +61-3-99059017

Citation: Berger, E.; Mackie, G.; Reupert, A.; Greenfeld, D.; Allen, K.-A.; May, F.; Wurf, G.; Summers, D.; Morris, Z. The Experiences of Australian School Mental Health Professionals during COVID-19 Lockdowns. *Children* **2023**, *10*, 1157. https://doi.org/10.3390/children10071157

Academic Editors: Matteo Alessio Chiappedi and António J. Santos

Received: 13 May 2023
Revised: 26 June 2023
Accepted: 30 June 2023
Published: 1 July 2023

Abstract: Young people have emerged as one of the most impacted groups from the COVID-19 pandemic and related restrictions to daily activities, with disruptions to schooling, social interactions, and connections. Simultaneously, students' access to school mental health professionals were restricted or modified. The aim of this paper was to identify how school mental health professionals supported and addressed the mental health needs of young people during COVID-19 restrictions in Australia. School mental health professionals were surveyed during the 2020 lockdowns using a questionnaire designed by researchers in the United States of America. The innovations school mental health staff adopted to support students during lockdowns and remote learning were presented, including telehealth services, digital resources, and the online training and support they received/provided. The barriers and facilitators to providing counselling and assessment services during lockdowns were identified, including issues with providing psychometric assessments during remote learning, and ethical concerns when delivering remote counselling to students. Recommendations have been included, which address how school mental health professionals could be supported to assess and treat young people during future pandemics and school restrictions.

Keywords: COVID-19; pandemic; students; mental health professional; psychologist; counsellor; therapy; assessment

1. Introduction

1.1. The COVID-19 Pandemic

On 11 March 2020, the novel coronavirus 2019 (SARS-CoV-2; COVID-19) was declared as a global pandemic [1]. People infected with COVID-19 may require hospitalisation due to severe respiratory failure, while others present with milder symptoms, including shortness of breath, fever, and/or fatigue. To prevent the spread of COVID-19 and reduce the rates of infection and death, social distancing and lockdown laws were introduced across many districts. Australia introduced some of the most restrictive lockdown laws and measures in the world, including the closure of businesses, restaurants, sporting venues, schools, and universities, as well as the introduction of mandatory facemasks and social distancing [2]. Although these measures have reduced the morbidity and mortality rates associated with COVID-19, researchers have recorded increased psychological distress, family violence rates, substance abuse, and suicidality during the pandemic [3].

1.2. Psychological Impacts of COVID-19

A recent systematic review found that due to the COVID-19, severe acute respiratory syndrome (SARS), and Middle East respiratory syndrome (MERS) outbreaks, adults have reported suffering from increased depression, anxiety, and stress [4]. Another review by Brooks and colleagues (2020) found post-traumatic stress, confusion, and anger for adults and children quarantined during the SARS, MERS, Ebola, and H1N1 influenza outbreaks [5]. When parents were asked about the psychological impacts of COVID-19 and lockdown restrictions on their children, they reported increased hyperactivity, restlessness, inattention, anger, anxiety, and irritability [6–8]. Australian adolescents have reported suffering from a poorer mental health (e.g., anxiety and depression) since the beginning of the pandemic, with concurrent negative impacts on peer relationships, learning outcomes, and family relationships [9]. Moreover, Australian adults and adolescents with pre-existing mental health conditions have experienced significantly greater negative emotions (e.g., anxiety and depression) and negative wellbeing outcomes (e.g., sleep disturbance, psychological distress, and loneliness) during the pandemic, relative to the general population [9,10].

Parents and educators have reported similar effects of COVID-19 on their emotional wellbeing and their subsequent capacity to support children and students [11–13]. In a study of teachers' descriptions of their work during the pandemic in Norway, Sweden, and the United States, COVID-19 school restrictions (e.g., online learning) were found to have caused significant worries about teachers' health, potential job losses, and pay cuts [14]. Teachers from Canada and the United States reported elevated levels of psychological distress associated with high levels of concern for student wellbeing and an increased awareness of inequities among their students [15,16]. Similarly, a review of the impact of COVID-19 on healthcare workers found consistent reports of stress, anxiety, and depressive symptoms because of COVID-19 [17]. These psychological symptoms were found to be the most prominent amongst nurses, female workers, front-line healthcare workers, younger medical staff, and workers in areas with high COVID-19 infection rates.

1.3. COVID-19 and School Mental Health Care

School-based mental health professionals have always played a key role in supporting young people's mental health and wellbeing within the school environment [18]. This has been achieved by providing both individual, student-level services, as well as broader system-level services [19]. Student-level services focus on evidence-based preventative and intervention strategies to manage student mental health concerns (e.g., counselling and therapies), skill building and development (e.g., social skills), assessment and testing (e.g., cognitive, and academic assessments), as well as vocational guidance (e.g., careers counselling) [19–21]. System-level services support young people through school-wide practices and policy development (e.g., developing anti-bullying policies), and by promoting the capacity of parents/caregivers and teachers to support young people [19]. In Australia, prior to the COVID-19 pandemic, most of these services were conducted in-person within the school environment.

Little is known about how school-based mental health professionals in Australia supported children and adolescents during the COVID-19 pandemic. To address a similar knowledge gap in the United States, Schaffer, and colleagues (2021) developed a 53-item survey that targeted full-time school psychologists employed in K-12 schools [22]. Based on the responses received, the authors found that psychologists faced difficulties, challenges, and changes to their roles and responsibilities, especially related to the types of services they offered during this period. While this study captured participants' concerns and insights into how school psychology was impacted by COVID-19, given that COVID-19 restrictions varied between countries, it is likely that the impact of the pandemic on school-based mental health professionals is country-specific. Therefore, this study aimed to identify the experiences and practices of school mental health professionals (e.g., school psychologists and counsellors) during these COVID-19 restrictions in Australia, including support for at-risk or vulnerable young people. There are indications that access to online and telephone counselling increased during the pandemic [23–25]; however, it is unclear what (if any) adaptations were made by school-based mental health practitioners in response to the pandemic, school closures, and online learning restrictions. Given there have been seven international health crises in the last 20 years [26], the outcomes of this research can be used to guide mental health service delivery in schools during other pandemic or disaster events also causing school closures and lockdowns. The research questions for this study included:

1. What online practices did Australian school-based mental health professionals employ to support young people's mental health during these COVID-19 restrictions, including identifying and supporting the students who were at risk?

2. What were the barriers and facilitators in addressing students' mental health needs during these COVID-19 restrictions in 2020?

3. What do school-based mental health professionals recommend in regard to assessment and treatment innovations for students during future pandemics or similar events?

2. Materials and Methods

2.1. Sampling and Recruitment

A purposive sampling method was used to recruit participants who worked as mental health professionals within Australian schools, including psychologists, counsellors, social workers, and wellbeing team members. After ethics approval from the Monash University Human Research Ethics Committee was provided, digital flyers and notices advertising the research project and a link to complete the survey were distributed amongst the personal networks of the research team via professional associations (e.g., the Australian Psychological Society), and through paid social media advertisements (including Facebook, LinkedIn, Instagram, and Twitter). Paid advertisements were targeted toward individuals based on their school-based mental health occupation Australia-wide.

2.2. Participants

A total of 105 participants responded to this survey. After data cleaning was completed, 20 respondents were subsequently removed due to either not providing consent to participate or providing no further data outside of consenting to participate. Therefore, the final sample consisted of 85 participants aged between 24 and 59 years old, respectively (M = 38.11 years, SD = 8.12 years). Of this sample, 92.69% of participants identified themselves as women and 7.32% identified as men. The majority of participants were registered as psychologists with the Psychology Board of Australia (75.61%), with counsellors (15.85%), social workers (3.66%), provisionally registered psychologists with the Psychology Board of Australia (3.66%), and wellbeing staff (2.44%) also being represented within this sample. Participants' years of experience in their various school-based roles ranged from 1 to 26 years (M = 6.44 years, SD = 6.46 years).

The majority of the sample reported working in either Victoria (55.41%) or New South Wales (33.78%), with Queensland (5.41%), Tasmania, (2.70%), Western Australia (1.35%), and the Australian Capital Territory (1.35%) also being represented in this sample. Most participants reported working in metropolitan schools (75.31%), while the rest worked in regional schools (19.75%) and rural/remote schools (4.94%). Over one quarter of participants (28.05%) reported being in Stage 4 lockdown at the time of completing the survey, with a further 15.85% in Stage 3, 23.17% in Stage 2, and 7.32% in Stage 1 (see below for a description of these stages). Furthermore, nearly one-quarter (23.17%) of participants were unsure of what COVID-19 lockdown restriction stage they were in, while a further 1.22% reported being under no restrictions. Most participants' primary location of work at the time of completing the survey was either at school (41.46%) or at home (40.24%), with a further 18.29% working in a combination of both home and school settings.

Restrictions and stages of lockdown varied between the different Australian states. Stage 1 restrictions included the ban of non-essential outdoor gatherings of 500 or more people, indoor gatherings of more than 100 people, and restricting visitors to aged care.

Stage 2 included closure of non-essential services, and people were encouraged to work from home. Stage 3 included people only being allowed to leave home for four reasons, including shopping for food and essential supplies, receiving medical care, exercising, and attending work. Moreover, teaching and learning were moved online for prep (first year of school) to Year 10 students for those under Stage 3. Years 11 and 12 students continued to attend classes at school, and vulnerable students (e.g., students with a disability) were able to be supervised at school. Stage 4 included further restrictions, such as people not being allowed to travel more than 5 km from home without a permit, a curfew from 8 pm to 5 am, and exercise in outdoor public spaces being limited to only one hour per day. At Stage 4, schools were closed to learning, with all teaching and learning activities for students from prep to Year 12 conducted online. However, vulnerable students were still able to be supervised in school.

2.3. Materials

Online Survey

The online survey used in the current study was initially developed by a United States-based research team [22] and was modified to suit the Australian context with permission from the authors. All participants were asked to base their responses on the delivery of school psychologist services at the peak of the pandemic, and to consider the impacts to their services during remote learning. Participant demographics were collected, which included age, gender identity, state or territory location of work, professional role within schools, years of experience in their current role, COVID-19 restrictions in place at their location, and primary location of work (i.e., school, home, or school and home) during the COVID-19 restrictions.

The detailed research questions and a review of peer-reviewed articles in the areas of crisis intervention, distance learning, tele-counselling, and previous pandemic impacts were all integrated and used to develop the survey items [22]. The survey sought to determine: (1) the general types of psychological support and services delivered to students during the COVID-19 restrictions (three items: one 'select all that apply' item and two open-ended items); (2) the delivery of counselling services to students during the COVID-19 restrictions (three open-ended items); and (3) the delivery of psychological assessments to students during the COVID-19 restrictions (three open-ended items). For more information about the survey, please refer to Reupert and colleagues (2021) and Schaffer et al. (2021) [22,27].

2.4. Procedure

After clicking on the link provided via the digital flyers, participants were presented with the explanatory statement to the study. A consent button was presented at the end of this document. Clicking on this button indicated informed consent to participate in the study. Participants were then presented with the online survey. The survey was open for participant completion from 18 May 2020 to 26 October 2020. Victorian schools were closed for most of May to October, New South Wales schools were closed from around April to June, and the other states experienced shorter school closure periods from around March to April 2020, respectively.

2.5. Data Analysis

Survey data were analysed using IBM Statistical Package for Social Sciences 26 (SPSS-26). Descriptive statistics were calculated for all the demographic variables of interest to our study, including frequencies and percentages for the categorical and dichotomous variables, and means and standard deviations for the continuous variables. The open-ended items were analysed using Braun and Clarke's (2006) six-phase approach to thematic analysis [28]. These phases include: (1) familiarisation with the data; (2) initial coding; (3) searching for themes; (4) reviewing themes; (5) defining and naming themes; and (6) finalising analysis write-up. Two authors independently analysed the open-ended data using thematic analysis. The themes developed by both authors were then compared, and a final set of themes was developed by the authors. The open-ended data were then coded to determine the number and percentage of participants who provided a response related to each theme. After coding, descriptive statistics were then used to determine the number and percentage of participants who responded to each theme.

3. Results

The results are presented under each of the three research questions for this study.

RQ 1: What online practices did Australian school-based mental health professionals employ to support young people's mental health during these COVID-19 restrictions, including identifying the students who were at risk?

Participants were asked to select the ways in which they had been delivering social-emotional, behavioural, or academic support services to students during the COVID-19 restrictions from the options listed in Table 1. Responses were received from 77 of the 85 participants and are summarised in Table 1.

Table 1. Social-emotional, behavioural, or academic online support services delivered to students amid COVID-19 restrictions.

Response	N (n = 77)	%
Telehealth interventions or tele-counselling via webcam or phone	66	86
Use of database such as Google classroom to post social, emotional, behavioural, or academic support	22	29
Developing or posting videos on common social, emotional, behavioural, or academic issues that parents/caregivers can use with/on their children (psychoeducation)	20	26
Mailing packets or newsletters with social, emotional, behavioural, or academic interventions	18	23
Other	16	21

Note: The percentage adds up to more than 100%, as participants may have selected more than one option.

As presented in Table 1, the most common social-emotional, behavioural, and academic online support services delivered to children during these COVID-19 restrictions were telehealth interventions or tele-counselling, with 86% of participants providing this service. The next most provided support, reported by 29% of participants, was the use of online databases to post/share social-emotional, behavioural, or academic support content with students. This was closely followed by sharing videos with parents/caregivers about the common social-emotional, behavioural, or academic issues faced by children (26%). Finally, 21% of participants utilised other support services delivered via emails to students and parents/caregivers through the provision of articles and resources posted on school websites or online wellbeing hubs, along with the upskilling of staff to respond to students' needs through the delivery of online professional development and participation in individual education plan (IEP) meetings.

Table 2 presents participants' responses to the open-ended question 'Have you and/or your school undertaken any novel strategies to support student wellbeing given COVID-19 restrictions? If yes, please specify'. Overall, 18 participants responded no, 13 selected unsure, 42 said yes, and 12 participants did not respond, respectively. Participants who responded with yes provided an open-ended response of the novel strategy they used to support student wellbeing.

Table 2. Novel strategies introduced to support student wellbeing.

Theme	N (n = 42)	%
Phone, email, and online support for students	42	100
Online staff training and support	7	17
Student wellbeing survey	4	10
Changes to student curriculum	5	12
Support for and engagement with parents/caregivers	3	7

Note: The percentage adds up to more than 100%, as participants may have reported more than one option.

As shown in Table 2, there were several novel strategies adopted by school mental health practitioners to support student wellbeing during the COVID-19 restrictions. The most common strategy adopted by participants was to provide telephone, email, and online support services for students. This included both group and individual student online support, resources, activities, and telephone calls regarding students' welfare and wellbeing, as well as emails sent to students. This was followed by the next most common strategy being adopted by 17% of participants, which was online staff training and support. Less frequently noted, but still noted as novel strategies, were changes to the student curriculum and workload (12%), the utilisation of student wellbeing surveys (10%), and support for and engagement with students' parents/caregivers (7%).

Participants were then asked, 'Have you and/or your school undertaken any novel strategies to support the identification of at-risk students? If yes, please specify'. Thirty-three participants responded no, 26 responded yes, 12 responded unsure, and 14 participants did not respond, respectively. The thematic analysis results from the 26 participants who responded yes to this question and provided an open-ended response are presented in Table 3.

Table 3. Strategies to identify at-risk students.

Theme	N (n = 26)	%
Survey and self-assessments for students	8	31
Contact with parents	8	31
Peer support	1	4
Whole school approaches	8	31
Monitoring attendance and engagement	6	23
Referral management	5	19
Other	4	15

Note: The percentage adds up to more than 100%, as participants may have reported more than one option.

Table 3 presents the novel strategies that were employed by school mental health practitioners to identify the at-risk students. The most common strategies included administering surveys and self-assessments to students (31%), contacting parents of students (31%), and developing and implementing whole school approaches (31%), such as consistent wellbeing policies and processes. These were followed by the next most utilised strategies, including monitoring student attendance and engagement with school (23%), managing the referrals of students (19%), and other strategies (15%), such as having teachers/mentors visit student homes to drop off schoolwork, liaising with year level coordinators, and encouraging peer-to-peer contact and support (4%).

RQ2: What were the barriers and facilitators in addressing students' mental health needs during these COVID-19 restrictions?

Table 4 presents the barriers that have challenged school mental health practitioners in conducting psychological assessments with students during the COVID-19 restrictions. Participants were asked 'Given COVID-19 restrictions, what are three barriers (if any) that have challenged you as a school psychologist/counsellor to assess students?' A total of 31 participants responded to this question. The most common barrier faced by 39% of participants was the inability to assess students face-to-face to accurately evaluate their psychological needs. The next most common barriers faced by 12% of participants were issues related to students' assessment in terms of adapting the content of the assessment to an online platform, lack of referrals for the assessments, and advice to cease assessments during the COVID-19 restrictions. Less frequently noted, but still identified as barriers to assessing students, were technical difficulties in performing assessments online (19%), lack of practitioner knowledge or ability to assess students online (16%), time constraints (13%), lack of parent collaboration (10%), issues with confidentiality/privacy (10%), and individual student factors (13%), such as lack of attendance and poor behaviour. Other barriers (26%) included staff lack of supervision, lack of clear directives for assessments, and all assessments ceasing during the lockdowns.

Table 4. Barriers to assessing students.

Theme	**N (*n* = 31)**	%
Time constraints	4	13
Confidentiality/privacy	3	10
Inability to assess face-to-face	12	39
Technical difficulties to perform assessment online	6	19
Lack of ability and knowledge to assess online	5	16
Issues with the content of the assessment and their adaptation to online mode	8	12
Interface with school	8	26
Parent collaboration	3	10
Individual student factors	4	13
Other	8	26

Note: The percentage adds up to more than 100%, as participants may have reported more than one option.

Participants also identified several facilitators to assessing students during the COVID-19 restrictions. Participants were asked 'Given COVID-19 restrictions, what are three things (if any) that have helped you as a school psychologist/counsellor to assess students?' Nineteen participants responded to this question. As presented in Table 5, the most common facilitators reported by 21% of participants were having access to peer support and supervision, as well as access to alternative forms of communication to contact students and parents (e.g., online platforms, telephones, and through the internet; 37%). The next most common facilitators reported by 16% of participants were collaboration with parents and having enough time to complete their assessments (16%). Less commonly reported facilitators were having access to technology, such as iPads for assessment purposes (11%), the implementation of workplace policies and procedures (11%), adequate or suitable workspaces to conduct assessments (11%), and other reasons (11%), such as collaboration with teams and supervisors. It should be noted that 16% of participants reported that they did not conduct assessments during the COVID-19 restrictions, and thus were unable to provide insight into facilitators. Additionally, a further 11% of participants did not identify any facilitators to the assessment of students.

Table 5. Facilitators to assessing students.

Theme	**N** **(*n* = 19)**	%
Access to alternative communication	7	37
Access to technology to perform assessments	2	11
Workplace policies and procedures	2	11
Adequate/suitable workspace	2	11
Collaboration with parents	3	16
Having enough time	3	16
Peer support/supervision	4	21
Did not conduct assessments during COVID-19	3	16
None	2	11
Other	2	11

Note: The percentage adds up to more than 100%, as participants may have reported more than one option.

Facilitators to provide counselling services are presented in Table 6. Participants were asked 'Given COVID-19 restrictions, what are three things (if any) that have helped you as a school psychologist/counsellor to counsel students?' Of the 85 participants, 40 provided a response to this question. The most common facilitator was access to online platforms, which was reported by 50% of participants. The next most common facilitator was having access to a supportive team within the school (38%), including other mental health practitioners, teachers, and employers. Following this, 23% of participants reported that having access to digital resources functioned as a facilitator in providing counselling services to students. Less commonly reported facilitators included access to supervision and peer consultation (18%) and other facilitators (13%), such as having an appropriate office space at home and having prior familiarity with online platforms.

Table 6. Facilitators to providing counselling to students.

Theme	**N** **(*n* = 40)**	%
Access to online platforms	20	50
Access to digital resources	9	23
Supportive team	15	38
Supervision and peer consultation	7	18
Professional development	10	25
Other	5	13

Note: The percentage adds up to more than 100%, as participants may have reported more than one option.

Participants were asked 'Given COVID-19 restrictions, what are three things (if any) that have challenged you as a school psychologist/counsellor to counsel students?' Of the 85 participants sampled in this study, 56 provided a response to this question. As presented in Table 7, the most common barriers to providing counselling to students during the COVID-19 restrictions were issues with student attention and engagement on online platforms, reported by 29% of participants. The next most common barrier, reported by 25% of participants, was issues with confidentiality/privacy of students due to being within the family home. This was followed by technology and internet connection issues, which was reported by 23% of participants. Less commonly reported barriers included lack of face-to-face interactions (20%) and other barriers (14%), such as client age and capacity to engage in online counselling.

Table 7. Barriers to providing counselling to students.

Theme	N (*n* = 56)	%
Technology and internet issues	13	23
Issues with student attention and engagement	16	29
Lack of face-to-face interactions	11	20
Issues with confidentiality/privacy	14	25
Other	8	14

Note: The percentage adds up to more than 100%, as participants may have reported more than one option.

RQ3: What do school-based mental health professionals recommend in regard to assessment and treatment innovations for students during future pandemics or similar events?

Participants were asked 'Given COVID-19 restrictions, what are three things (if any) your school could do to support you to assess students?' Overall, 32 participants provided a response to this question. As presented in Table 8, the most common recommendations of support schools could provide to enable school mental health practitioners to assess students was the access to suitable and relevant resources (25%), such as appropriate technology and forms of communication. The next most common recommendations of support were being provided with access to additional support (19%), such as peer consultation and supervision, and being provided with funded resources (16%) to assist with assessments. Less frequently reported recommendations of support were being provided with funded professional development (6%), schools being open to change and flexibility regarding the administration of assessments (6%) and developing policies and procedures to support students in novel ways (3%). Thirteen percent of participants reported that they already felt supported by their schools to assess students, while some participants (13%) reported not being sure of how the school could support assessments with students.

Table 8. Recommended support to assess students.

Theme	N (*n* = 32)	%
Access to resources	8	25
Funded professional development	2	6
Funded resources	5	16
Additional support	6	19
Openness to change/flexibility	2	6
Policies and procedures to support students in new ways	1	3
Already supported	4	13
Other	4	13

Note: The percentage adds up to more than 100%, as participants may have reported more than one option.

Participants were then asked 'Given COVID-19 restrictions, what are three things (if any) your employer could do to support you to counsel students?' Thirty-nine participants responded to this question. As presented in Table 9, the most common recommendation of support employers was increased wellbeing and support offered to school mental health practitioners (21%). This included regular staff wellbeing check-ins and recognising staff burnout. The next most common recommendations of employer support were increased role flexibility (15%) and establishing better referral processes (15%). Thirteen percent of participants already felt supported by their employers, while 10% reported that nothing could be recommended as the barriers were outside of their employers' control (e.g., childcare access for their own children). A smaller percentage of participants recommended access to IT support (10%), access to supervision (10%), and access to training in telehealth

(5%). Thirteen percent of participants reported other recommendations, such as a more consistent communication between staff and assistance with interpreters for parents.

Table 9. Recommended support to counsel students.

Theme	N (n = 39)	%
IT support	4	10
Training in telehealth	2	5
Increased flexibility	6	15
Increased staff wellbeing and support	8	21
Provide supervision	4	10
Better referral process	6	15
Already felt supported	5	13
Nothing	4	10
Other	5	13

Note: The percentage adds up to more than 100%, as participants may have reported more than one option.

4. Discussion

This study explored the experiences and practices of Australian school mental health professionals during the COVID-19-related lockdowns and school closures. Use of telehealth/tele-counselling interventions to replace face-to-face counselling was a significant change made by participants. Participants also reported providing more online support and professional development to their colleagues during these lockdowns. Others have highlighted an increase in online counselling services (e.g., tele-counselling, online group intervention/workshops, and telephone sessions), as well as online support/resources (e.g., communication media) in international educational settings during these COVID-19 lockdowns and/or restrictions [29,30]. Schaffer et al. (2021), as well as Reupert et al. (2021) reported comparable results, which confirmed that an increasing number of psychologists shifted to remote delivery methods to ensure the continuation of psychological services [22,27].

While there are many advantages and strengths to online counselling and telehealth services (e.g., ability to socially distance, and reduced travel time) [31], researchers have also reported on various issues for professional and ethical practice associated with online counselling services. Common issues include risks to client/patient privacy and confidentiality (e.g., cyber-security), matching the mode of care to the client/patient (e.g., some health issues are less suitable for online services), health professional competency to deliver services using an online platform/system, and the ability to respond to the risk or emergency issues (e.g., disclosure of suicidal thoughts or intentions) [32,33]. The results from the current study supported these findings within the context of schools and working with young people, highlighting that issues of privacy, confidentiality, and online competency were also experienced within the telehealth services offered in schools.

The adoption of remote practices for students deemed to be at risk of developing severe mental health conditions was another innovation reported by the participants. This included the use of online surveys and self-assessment tools, greater contact with parents of at-risk students, and adoption of new policies to manage the high needs of these students. Reflecting on the ethical and professional practice standards of remote mental health service delivery for vulnerable students, it is crucial for school mental health professionals to be aware of the potential practice and ethical challenges that may arise from this work [34]. For instance, privacy and confidentiality may be breached when attending counselling from within the home/family environment, which may work against at-risk students engaging with the online counselling services. Long standing research has shown that adolescents in general are reluctant to access mental health services due to barriers, such as stigma and confidentiality concerns [35,36]. The impact of tele-counselling on students' help-seeking intentions and behaviour was outside of the scope of this study. However, reports that mental health providers were concerned about confidentiality with students may have exacerbated students' reluctance to seek help during the pandemic.

Other ethical and professional issues were also noted in the results of the current study. Participants reported that a common barrier to supporting students during the pandemic was their inability to provide face-to-face services. One of the services provided by school mental health staff that was most severely impacted by the pandemic was their ability to conduct psychological assessments with students. Participants reported that several assessments ceased during the pandemic. Research and standards are lacking for practitioners who are seeking to improve their skills to conduct online assessments with clients. Krach and colleagues (2020) found that assessment publishers acted quickly to offer computer-adapted, tele-assessment methods; however, these methods were developed using minimal empirical data, and test publishers and government agencies often recommended using tele-assessments methods with caution [37]. Additionally, other studies have highlighted the issues associated with inadequate training for school mental health professionals to use and access tele-assessments [38,39], suggesting that further professional training and development is required to ensure that school mental health professionals are able to competently use online assessment methods.

Our results suggest there is still significant work to be completed to improve the confidence, knowledge, and skills of the practitioners delivering tele-assessment tools. Professional development and protocols are needed to assist mental health professionals to feel confident and equipped to administer, interpret, and make recommendations for students' mental health based on online assessments. It is possible that tertiary students, who completed their mental health training and practicums during the pandemic, will be more confident and prepared to deliver and interpret online psychological assessments compared to experienced practitioners, who have limited experience with these methods of assessments. However, more research is needed to explore the aspects of validation, interpretation, and reporting of psychological assessment tools delivered online [38].

While participants noted issues concerning the conduct of online assessments during the pandemic, and a lack of confidence delivering online assessments, participants acknowledged that colleagues provided a great deal of support around the remote delivery of assessments and counselling during the pandemic. Online platforms were also deemed to be more useful for participants in terms of providing counselling services to students during the pandemic and school closures compared to administering assessments. This study found that it was much easier for practitioners to shift to an online delivery mode of counselling compared to the online administration of assessments during the pandemic. However, consistent with the prior literature, the challenges associated with maintaining client confidentiality/privacy, matching online services to the client's needs, making sure professionals are competent in their use of online services, and developing protocols for responding to risk or emergency situations [34,38,40] were all raised by practitioners in Australia.

To address these issues, the Australian Psychology Society (APS) developed recommendations for providers of telehealth measures under the Better Access initiative. These recommendations provide information and resources to assist psychologists in safely and adequately utilising telehealth measures, such as implementing online security measures, assessing client suitability for online services, completing risk assessment and management, training in the use of telehealth, and adjustments to the informed consent and referral policies [41]. While these considerations are relevant in addressing some of the issues that participants faced when delivering online counselling services, they fail to address the ethical and practical issues that are associated with online assessment administration. This suggests that Australian school psychologists still lack the appropriate guidance and resources to assist the administration of online assessment services. It is reasonable to assume that restrictions and school lockdowns will occur in the future in response to COVID-19, other pandemic events, or other disaster events [26]. Thus, continued research and practice protocols concerning the appropriate and evidence-based administration of assessment tools during lockdown periods are required. Research on the impacts for students, schools, and families on the lack of access to psychological assessments over the course of Australia's 2020 and 2021 lockdowns is also needed.

These recommended considerations from the APS (2022) are largely consistent with the recommendations offered by participants in the current study in terms of the delivery of remote assessment/counselling for students [38]. For example, in line with the APS document, many participants recommended better telehealth training, adjusting referral policies and procedures for students, better IT support, and more general resources for telehealth services. Participants also recommended an additional focus on staff wellbeing and support, as well as additional supervision, which were not specifically outlined by the APS. Thus, in order for school psychologists to continue providing remote services to students, it will be important for schools to ensure that school mental health practitioners are well supported both personally and professionally during future lockdown or remote working circumstances.

5. Limitations

This study is the first known study to explore the experiences of school mental health staff in response to students during the COVID-19 pandemic in Australia. However, the small sample size limits the generalisability of the results. Unfortunately, there is a lack of known data on the gender and occupation distribution of school mental health professionals in Australia. However, data on the gender distribution of psychologists in Australia, working across a variety of a settings, shows that our data were skewed towards female rather than male participants. Data were also skewed towards the Victorian and New South Wales-based mental health providers; however, national data shows that Victoria and New South Wales have the highest number of registered psychologists compared to the other states [42]. The open-ended questions used in the survey were able to capture the perceptions and experiences of school mental health practitioners; however, these questions increased the length of the overall survey. It is likely that the additional time and effort required of participants to respond to these questions resulted in some participants choosing to not provide a response. Furthermore, analysing qualitative responses through a questionnaire did not permit probing and more detailed responses as would be the case for an interview or focus group data. For example, a few participants stated they had not adopted novel strategies to support students, including at-risk students, during the pandemic, and it was unclear from their responses why this was the case. The qualitative data captured also did not permit statistical comparisons between practitioners from the different states and regions in Australia, between mental health practitioners who have distinct roles in schools, or staff from different education sectors (e.g., primary compared to secondary school). The results of this study indicated that older students were more able to engage with practitioners using telehealth and other remote services compared to the younger students. Therefore, the experiences of professionals in secondary schools may be different to those working in primary schools. The different lockdown and school closure restrictions imposed by these different states also potentially impacted the comments and expertise of participants. Mental health practitioners from Victoria, which experienced the longest lockdowns in Australia, were deemed to be more likely to have developed greater expertise and adopt more remote online practices during the 2020 lockdowns, compared to professionals from other states.

6. Conclusions

The current study was the first known piece of empirical research to explore how school mental health professionals addressed and supported the mental health needs of young people during the COVID-19 restrictions in Australia. These findings demonstrated that school mental health professionals relied upon and adopted online and telehealth services to support young people during the COVID-19 restrictions. Whilst access to online platforms/services, additional staff support, and professional supervision were amongst the facilitators to supporting young people's mental health needs, issues, such as technology/internet difficulties, confidentiality and privacy concerns, and online student engagement were all found to act as barriers to supporting young people. The concerns raised by school mental health providers should be considered when developing future resources and professional learning activities to ensure that suitable and evidence-based psychological support can be provided to school students.

Author Contributions: Conceptualization, E.B., G.M., A.R., K.-A.A., F.M., G.W., D.S. and Z.M.; Methodology, E.B., G.M., A.R. and D.G.; Validation, E.B., G.M. and D.G.; Formal Analysis, G.M. and D.G.; Investigation, E.B., G.M., A.R. and D.G.; Resources, A.R.; Data Curation, All authors; Writing—Original Draft Preparation, E.B., G.M. and D.G.; Writing—Review and Editing, All authors; Supervision, E.B.; Project Administration, E.B. and A.R.; Funding Acquisition, A.R. All authors have read and agreed to the published version of the manuscript.

Funding: This research received no external funding.

Institutional Review Board Statement: Ethical Approval was provided by Monash University Human Research Ethics Committee, Approval number: 24559, Date: 16 August 2021.

Informed Consent Statement: Informed consent was obtained from all subjects involved in the study.

Data Availability Statement: Not available.

Conflicts of Interest: The authors declare no conflict of interest.

References

1. World Health Organisation. WHO Director-General's Opening Remarks at the Media Briefing on COVID-19—11 March 2020. 2020. Available online: https://www.who.int/director-general/speeches/detail/who-director-general-s-opening-remarks-at-the-media-briefing-on-covid-19{-}{-}-11-march-2020 (accessed on 21 February 2022).
2. Berger, E.; Reupert, A. The COVID-19 pandemic in Australia: Lessons learnt. *Psychol. Trauma: Theory Res. Pract. Policy* **2020**, *12*, 494–496. [CrossRef] [PubMed]
3. Humphreys, K.L.; Myint, M.T.; Zeanah, C.H. Increased risk for family violence during the COVID-19 Pandemic. *Pediatrics* **2020**, *146*, e20200982. [CrossRef] [PubMed]
4. Berger, E.; Jamshidi, N.; Reupert, A. Psychological impacts of coronavirus outbreaks on adults: A rapid evidence review. *Int. J. Ment. Health Promot.* **2022**, *24*, 619–634. [CrossRef]
5. Brooks, S.K.; Webster, R.K.; Smith, L.E.; Woodland, L.; Wessely, S.; Greenberg, N.; Rubin, G.J. The psychological impact of quarantine and how to reduce it: Rapid review of the evidence. *Lancet* **2020**, *395*, 912–920. [CrossRef]
6. Di Giorgio, E.; Di Riso, D.; Mioni, G.; Cellini, N. The interplay between mothers' and children behavioural and psychological factors during COVID-19: An Italian study. *Eur. Child Adolesc. Psychiatry* **2020**, *30*, 1401–1412. [CrossRef]
7. Orgilés, M.; Morales, A.; Delvecchio, E.; Mazzeschi, C.; Espada, J.P. Immediate psychological effects of the COVID-19 quarantine in youth from Italy and Spain. *Front. Psychol.* **2020**, *11*, 579038. [CrossRef]
8. Romero, E.; López-Romero, L.; Domínguez-Álvarez, B.; Villar, P.; Gómez-Fraguela, J.A. Testing the effects of COVID-19 confinement in Spanish children: The role of parents' distress, emotional problems and specific parenting. *PsyArXiv Prepr.* **2020**, 6975. [CrossRef]
9. Li, S.H.; Beames, J.R.; Newby, J.M.; Maston, K.; Christensen, H.; Werner-Seidler, A. The impact of COVID-19 on the lives and mental health of Australian adolescents. *Eur. Child Adolesc. Psychiatry* **2021**, *31*, 1465–1477. [CrossRef]
10. Rossell, S.L.; Neill, E.; Phillipou, A.; Tan, E.J.; Toh, W.L.; Can Rheenen, T.E.; Meyer, D. An overview of the current mental health in the general population of Australia during the COVID-19 pandemic: Results from the COLLATE project. *Psychiatry Res.* **2021**, *296*, 2–8. [CrossRef]
11. Brown, S.; Doom, J.; Lechuga-Peña, S.; Enos Watamura, S.; Koppels, T. Stress and parenting during the global COVID-19 pandemic. *Child Abus. Negl.* **2021**, *110*, 104699. [CrossRef]

12. Prime, H.; Wade, M.; Browne, D.T. Risk and resilience in family well-being during the COVID-19 pandemic. *Am. Psychol.* **2020**, *75*, 631–643. [CrossRef]
13. Spinelli, M.; Lionetti, F.; Pastore, M.; Fasolo, M. Parents' Stress and Children's Psychological Problems in Families Facing the COVID-19 Outbreak in Italy. *Front. Psychol.* **2020**, *11*, 1713. [CrossRef] [PubMed]
14. Pramling Samuelsson, I.; Wagner, J.T.; Ødegaard, E.E. The Coronavirus Pandemic and Lessons Learned in Preschools in Norway, Sweden and the United States: OMEP Policy Forum. *Int. J. Early Child.* **2020**, *52*, 129–144. [CrossRef]
15. Baker, C.N.; Peele, H.; Daniels, M.; Samba, M.; Whalen, K.; Overstreet, S. The experience of COVID-19 and its impact on teachers' mental health, coping, and teaching. *Sch. Psychol. Rev.* **2021**, *50*, 491–504. [CrossRef]
16. Sokal, L.; Eblie Trudel, L. How to Prevent Teacher Burnout during the Coronavirus Pandemic, The Conversation. 2020. Available online: https://theconversation.com/how-to-prevent-teacher-burnout-during-the-coronavirus-pandemic-139353 (accessed on 21 February 2022).
17. Vizheh, M.; Qorbani, M.; Arzaghi, S.M.; Muhidin, S.; Javanmard, Z.; Esmaeili, M. The mental health of healthcare workers in the COVID-19 pandemic: A systematic review. *J. Diabetes Metab. Disord.* **2020**, *19*, 1967–1978. [CrossRef]
18. Brown, C.; Dahlbeck, D.T.; Sparkman-Barnes, L. Collaborative relationships: School counsellors and non-school mental health professionals working together to improve the mental health needs of students. *Prof. Sch. Couns.* **2006**, *9*, 332–335. [CrossRef]
19. Splett, J.W.; Fowler, J.; Weist, M.D.; McDaniel, H. The critical role of school psychology in the school mental health movement. *Psychol. Sch.* **2013**, *50*, 254–258. [CrossRef]
20. Smith, C.K.; Wood, S.M. Supporting the career development of gifted students: New role and function for school psychologists. *Psychol. Schs.* **2020**, *57*, 1558–1568. [CrossRef]
21. Wiener, J. The role of school psychologists in supporting adolescents with ADHD. *Can. J. Sch. Psychol.* **2020**, *35*, 299–310. [CrossRef]
22. Schaffer, G.E.; Power, E.M.; Fisk, A.K.; Trolian, T.L. Beyond the four walls: The evolution of school psychological services during the COVID-19 outbreak. *Psychol. Sch.* **2021**, *58*, 1246–1265. [CrossRef]
23. Headspace. Headspace Staff Experience of Telehealth during COVID-19. 2020. Available online: https://headspace.org.au/assets/Uploads/Telehealth-Staff-Experience-FINAL-8-10-20.pdf (accessed on 21 February 2022).
24. Jayawardana, D.; Gannon, B. Use of telehealth mental health services during the COVID-19 pandemic. *Aust. Health Rev.* **2021**, *45*, 442–446. [CrossRef]
25. Lifeline. Australians Reaching Out for Help in Record Numbers. 2021. Available online: https://www.lifeline.org.au/media/0olpjjfq/20210820-australians-reaching-out-for-help-in-record-numbers.pdf (accessed on 21 February 2022).
26. Reupert, A. Change and (the need for) adaptability: The new normal. *Adv. Ment. Health* **2020**, *18*, 91–93. [CrossRef]
27. Reupert, A.; Schaffer, G.E.; Von Hagen, A.; Allen, K.-A.; Berger, E.; Buttner, G.; Power, E.M.; Morris, Z.; Paradis, P.; Fisk, A.K.; et al. The practices of psychologists working in schools during COVID-19: A multi-country investigation. *Sch. Psychol.* **2021**, *36*, 190–201. [CrossRef]
28. Braun, V.; Clarke, V. Using thematic analysis in psychology. *Qual. Res. Psychol.* **2006**, *3*, 77–101. [CrossRef]
29. Konig, J.; Jager-Biela, D.J.; Glutsch, N. Adapting to online teaching during COVID-19 school closure: Teacher education and teacher competence effects among early career teachers in Germany. *Eur. J. Teach. Educ.* **2020**, *43*, 608–622. [CrossRef]
30. Savarese, G.; Curcio, L.; D'Elia, D.; Fasano, O.; Pecoraro, N. Online university counselling services and psychological problems among Italian students in lockdown due to COVID-19. *Healthcare* **2020**, *8*, 440. [CrossRef] [PubMed]
31. Situmorang, D.D.B. Online/Cyber Counseling Services in the COVID-19 Outbreak: Are They Really New? *J. Pastor. Care Couns.* **2020**, *74*, 166–174. [CrossRef]
32. Kramer, G.M.; Luxton, D.D. Telemental health for children and adolescents: An overview of legal, regulatory, and risk management issues. *J. Child Adolesc. Psychopharmacol.* **2016**, *26*, 198–203. [CrossRef]
33. Nelson, E.L.; Cain, S.; Sharp, S. Considerations for conducting telemental health with children and adolescents. *Child Adolesc. Psychiatr. Clin.* **2017**, *26*, 77–91. [CrossRef] [PubMed]
34. Chaet, D.; Clearfield, R.; Sabin, J.E.; Skimming, K. Ethical practice in telehealth and telemedicine. *J. Gen. Intern. Med.* **2017**, *32*, 1136–1140. [CrossRef]
35. Gulliver, A.; Griffiths, K.M.; Christensen, H. Perceived barriers and facilitators to mental health help-seeking in young people: A systematic review. *BMC Psychiatry* **2010**, *10*, 113. [CrossRef] [PubMed]
36. Rickwood, D.; Deane, F.P.; Wilson, C.J.; Ciarrochi, J. Young people's help-seeking for mental health problems. *Aust. E-J. Adv. Ment. Health* **2005**, *4*, 218–251. [CrossRef]
37. Krach, S.K.; Paskiewiscz, T.L.; Monk, M.M. Testing out children when the world shuts down: Analysing recommendations for adapted tele-assessment during COVID-19. *J. Psychoeduc. Assess.* **2020**, *38*, 923–941. [CrossRef]
38. Farmer, R.L.; McGill, R.J.; Dombrowski, S.C.; Benson, N.F.; Smith-Kellen, S.; Lockwood, A.B.; Powell, S.; Pynn, C.; Stinnett, T.A. Conducting psychoeducational assessments during the COVID-19 crisis: The danger of good intentions. *Contemp. Sch. Psychol.* **2021**, *25*, 27–32. [CrossRef] [PubMed]
39. National Association of School Psychologists. *Guidance for Delivery of School Psychological Telehealth*; National Association of School Psychologists: Bethesda, MD, USA, 2017.
40. Stoll, J.; Muller, J.A.; Trachsel, M. Ethical issues in online psychotherapy: A narrative review. *Front. Psychiatry* **2020**, *10*, 993. [CrossRef] [PubMed]

41. Australian Psychology Society. Telehealth Considerations for Providers. 2022. Available online: https://psychology.org.au/psychology/medicare-rebates-psychological-services/medicare-faqs-for-the-public/telehealth-services/telehealth-considerations-for-providers (accessed on 31 May 2023).
42. Psychology Board of Australia. Statistics. 2023. Available online: https://www.psychologyboard.gov.au/about/statistics.aspx (accessed on 31 May 2023).

children

MDPI

Article

School Burnout after COVID-19, Prevalence and Role of Different Risk and Protective Factors in Preteen Students

Noémie Lacombe *, Maryelle Hey, Verena Hofmann, Céline Pagnotta and Myriam Squillaci

Department of Special Needs Education, University of Fribourg, St-Pierre Canisius 21, 1700 Fribourg, Switzerland
* Correspondence: noemie.lacombe@unifr.ch

Abstract: Background: Current data show an increase in stress among youth since the COVID-19 pandemic, raising the question of the measures to be put in place to limit it. Aim: The aim of this study is to measure the prevalence of burnout and the different risk and protective factors of burnout among students and to compare the mean scores obtained with those collected in a similar sample in 2014. Method: Perceived health, school burnout, and the different risk and protective factors among students were measured by self-reported questionnaires among a sample of 184 students (11–13 years old). Results: The results indicate significantly higher burnout scores than in 2014. Different variables were predictive of school burnout: 1. At the individual level, a low academic achievement in mathematics, somatic symptoms, and the expressed stress; 2. At the interpersonal level, lack of support from peers, teachers, and parents and a negative classroom climate; 3. At the pandemic level, the increased workload related to the consequences of COVID-19. The factors with the greatest impact are confidence in the future, perceived stress, parental supports, and mathematics results. Conclusions: An intervention program targeting these four factors among burned out students would be relevant to set up in order to reduce its prevalence.

Keywords: school burnout; stress; risk factor; protective factors; COVID-19

Citation: Lacombe, N.; Hey, M.; Hofmann, V.; Pagnotta, C.; Squillaci, M. School Burnout after COVID-19, Prevalence and Role of Different Risk and Protective Factors in Preteen Students. *Children* **2023**, *10*, 823. https://doi.org/10.3390/children10050823

Academic Editor: Matteo Alessio Chiappedi

Received: 28 March 2023
Revised: 21 April 2023
Accepted: 26 April 2023
Published: 30 April 2023

1. Introduction

While burnout was initially measured in the adult professional world, the similarities between school and professional environments have prompted researchers to look at student burnout [1]. Indeed, research in the field has shown that the psychological manifestations of burnout in students are similar to those encountered in adults. These data are not surprising in themselves, as school is at the heart of children's lives, with a large number of hours, tasks, demands, and cognitive, instrumental, and social expectations, resulting in many potential stressors that students must face [2]. When demands exceed available resources, stress can get out of hand and lead to burnout [3]. According to Salmela-Aro et al. [1], this syndrome manifests itself in three dimensions: emotional exhaustion, cynicism about school, and feelings of inadequacy, which are close to the three-dimensional model of burnout developed by Maslach and Jackson [4] for professionals. More specifically, emotional exhaustion manifests itself as chronic fatigue linked to an overload of work, such as excessive school demands, perceived as uncontrollable [5,6]. Cynicism is defined as a loss of interest and meaning in schooling and related activities. The student may display a detached and indifferent attitude towards school. Finally, the feeling of inadequacy refers to a lack of achievement at school, a feeling of not 'fitting in' in the role of a student [6,7].

Table 1 provides a summary of the reported prevalence of school burnout found in 11 studies focused on school burnout. According to these data, the average prevalence stands at 14.36%. Among authors who measure the prevalence of burnout using the SBI, Gabola et al. [8] report 14.6% burnout among students aged 14.8 years on average. Among slightly younger Swiss students (11–13 years), Lacombe and Squillaci [9]. find that 4.5% of students would experience a high level of burnout. The authors justify these low results by the fact that the sample was younger than in the reference study [1]. Salmela-Aro and Tynkkynen [10] reach similar results: between 6.9% and 10.3% among 15-year-old students and between 9.9 and 16.9% among 17–18-year-old students. The same study distinguishes between the prevalence of burnout among young people in academic and vocational streams. According to their results, burnout is more frequent in academic streams.

Table 1. Prevalence of school burnout.

First Author	Country	N (% Girls), $M_{\hat{a}ge}$	Scale	Prevalence
Salmela-Aro (2012) [10]	FIN	687 (47.6%) *M* = 15, T1	SBI	8.6%
	FIN	749 (49.13%) *M* = 17.5, T4	SBI	12.5%
Lacombe (2015) [9]	SUI	313 (50.48%) Age: 11–13	SBI	4.5%
Kinnunen (2016) [11]	EUR	10,325 (52.4%) *M* = 15.19	SSBI	18.1%
Virtanen (2016) [12]	FIN	2485 (52.1%) *M* = 14.71	BBI-10	5.5%
Gerber (2018) [13]	SUI	249 (64.26%) [1] *M* = 16.09	SMBM	12%
	SUI	144 (32.64%) [2] *M* = 16.22	SMBM	6.9%
Lee (2019) [14]	KOR	1015 (57.3%) Age: 17–19	KABI	15.47%
Cheung (2019) [15]	CHN	1209 (39.8%) *M* = 14.85	MBI-SS	32.2%
Gabola (2021) [8]	SUI	343 (49.27%) *M* = 14.81, S1	SBI	14.6%
	ITA	497 (48.29%) *M* = 15.09, S2	SBI	27.6%
Summary	*K* = 11	*N* = 18,016		14.36%

Note. SUI = Switzerland; ITA = Italy; FIN = Finland; EUR = Europe, CHN = China; KOR = Korea. [1] Academic track; [2] Vocational track.

The Italian sample of Gabola et al. [8] contains 27.6% of burnout students. Compared to the Swiss sample of the same study (14.6%), the authors highlight the significant effect of age and nationality on the results. They hypothesize that Italian students do not have access to the same health services as Swiss students. Another hypothesis could be put forward, related to the COVID-19 pandemic. As the data for this study were collected during the year 2020 and the health situation in Italy was particularly critical at the beginning of the pandemic [16] it may have influenced the results.

Regarding studies, in which other measurement instruments than the SBI were used, prevalences relatively similar to the 10–15% stated by [1] were found in particular: 18.1% measured by the shortened version of the SBI, the Short Student Burnout Inventory (SSBI) [11] 6.9% and 12% measured by the SMBM [13] or 15.47% by the KABI [14].

In their Finnish study using the BBI-10, Virtanen et al. [12] surveyed 2485 students with an average age of 14.7 years. They report that 5.5% of their sample, or about 137 students, have a low engagement profile, coupled with high burnout. In addition, 40.6% of the students would correspond to a high engagement and low burnout profile, while 53.9% would have a medium level of engagement and burnout. The authors do not comment on this low result.

As for Cheung and Li [8], despite a moderate level of academic burnout in the overall sample, they found 32.2% (N = 389) of students in the burnout profile. These students reported high levels of exhaustion and cynicism and moderate levels of work inefficiency. Moreover, 57.8% of the sample fell into the 'moderately engaged' profile, while only 10% fell into the 'high functioning' profile. These prevalences, collected through the MBI-SS, were measured in a sample of Chinese students. The authors were alarmed by the high percentage found in their sample. For them, the fact that academic performance is a top priority in their country explains the fact that these adolescents, although exhausted, continue their studies and are therefore so represented in the burnout profile.

Thus, among these results and despite the different tools used, seven results out of eleven report prevalences between 7 and 18% of burnt-out students. Two results found lower prevalences and two studies higher prevalences.

To attempt to control the extent of the phenomenon, researchers investigate the risk factors of burnout, classifying them into three distinct categories [5,17,18]: individual, interpersonal, and organizational risk factors. Among the individual factors, gender and educational level are the most frequently studied variables. In several studies, girls have higher burnout scores than boys, especially in the dimensions of exhaustion and inadequacy [1,19]. However, other studies do not find significant differences between girls and boys [9,12,20,21]. As for the level of education, the results of studies show that students with a low level of education and/or learning difficulties are more likely to develop burnout than students who do not [22,23]. Moreover, current evidence shows negative correlations between academic achievement and school burnout [24,25].

Among the interpersonal factors, different variables have been identified in the research, among which peer, teacher, and parental support are central. Virtanen et al. [12] note that peer support is a protective factor against burnout, especially if it comes from peers with low levels of burnout [24]. In terms of teacher–student relationships, the more positive they are perceived to be, the better the performance and well-being of students [26,27]. A positively perceived school climate is a protective factor against burnout [23] whereas a school climate that is perceived negatively by students influences the severity of burnout [1].

Finally, parental support is also considered a protective factor against psychosocial health difficulties [23,28]. According to Meylan et al. [7], students who feel emotionally supported by their parents are less stressed by the demands of school and are less prone to burnout.

At the organizational level, research has identified various variables related to the organization of school life. According to Meylan et al. [20] students in academic tracks have higher burnout and exhaustion scores than students in vocational tracks. Salmela-Aro and Tynkkynen [10] and Bask and Salmela-Aro [29] also find this difference between the two study paths. Besides these individual, interpersonal, and organizational factors, other factors related to the pandemic must also be taken into account. Since 2020, the consequences of the COVID-19 on the health of children, adolescents, adults, and the elderly have brought to light new risk factors related to both physical and psychological health [30–35]. In this regard, a recent meta-analysis of 29 studies found that following the first year of COVID-19, 25.2% of young people showed elevated symptoms of depression and 20.5% elevated symptoms of anxiety [36]. The study by Tang et al. [35]) conducted among a sample of young people aged 6 to 17 years ($M = 11.86$, SD = 2.32), interviewed two months after the start of the pandemic, shows that 11.5% of young people presented clinical thresholds for anxiety, depression, and stress above the usual thresholds. Current data suggest that these figures have therefore doubled since the pandemic [36].

The present study aims to measure the burnout of 8th grade students one year after the lockdown in order to investigate the medium-term influence of the pandemic on their perceived health. The study follows on from research conducted in 2014 by Lacombe and Squillaci [9] which assessed the burnout in a sample of students with similar characteristics (age, school environment, Swiss canton). Thus, the mean scores in the three dimensions obtained in 2022 (T2) will be compared to those obtained in 2014 (T1) in order to check whether there is a difference in the perceived level of burnout. In addition, the study aims to identify risk factors related to the three dimensions of burnout.

This research answers the following three questions:

- Q1: What is the prevalence of school burnout among 8th grade pupils in the French-speaking part of Switzerland, in the canton of Fribourg? The hypothesis in line with the post COVID-19 student health reports [30] stipulates an increase in stress and depressive symptoms, which suggests an increase in social burnout.
- Q2: Are the mean scores in the three burnout dimensions different between T2 and T1? In line with studies on school burnout [9,12] among 12 to 16 year olds which report low burnout scores across their samples, the hypothesis is that even if the scores increase due to COVID, the score will remain correct when considering the whole sample.
- Q3: What are the personal, interpersonal, organizational, and COVID-19 pandemic-related factors that explain student burnout? According to previous research results, the following hypotheses can be formulated: (a) gender has no influence on school burnout [9,15]; (b) Low school performance increases the risk of burnout [3,9,11,37]; (c) Support from parents, friends, teachers reduces burnout [6,9,12]; d) Enjoying the classroom (perceived school climate) reduces the risk of school burnout [9].

2. Materials and Methods

Authorization to conduct the research was granted by the Directorate of Education and Cultural Affairs of the Canton of Fribourg (DFAC) in 2014 and in 2022. In 2014, 333 students in the canton of Fribourg were surveyed. The pupils filled in the paper questionnaires in the presence of the researcher. No personal data (first name, surname, photo, or video) was collected. For more details on the 2014 procedure, see Lacombe and Squillaci [9]. In 2022, the DFAC designated six institutions in which to conduct the research. After having contacted the schools' management, direct contact was established with each teacher in order to organize the anonymous completion of the questionnaires. As the participants are minors, permission forms were completed by the parents.

The sample comprised 184 preteens (11 to 13 years old; $M = 12.38$), including 98 boys and 86 girls. The participants came from ten classes of 8ème in the canton of Fribourg in Switzerland. In the canton of Fribourg, the school system consists of 8 years of elementary school (from 1st to 8th grade), concerning children from 4 to 12 years old. Then, schooling includes 3 years of secondary school (9th to 11th). Pupils then have the possibility of continuing their studies to obtain a high school diploma or to take up an internship in the professional world. The choice of pre-adolescents is justified by the lack of research on burnout scores in this sample. Indeed, many studies measure burnout in adolescents (13–18 years) see Table 1. However, it is crucial to measure school burnout in pre-adolescents in order to understand the timing of its appearance in the classroom. The sample included 78.3% students with a Swiss nationality (n = 144). Of the 21.7% foreign of participants, 12% (n = 22) were from other European countries, 4.9% (n = 9) from African countries, and 1.6% (n = 3) from Asian countries.

2.1. Materials and Procedure

Data collection took place in May 2022 approximately two years after the end of the pandemic lockdown. Two members of the research team administered the questionnaires to the students. The questionnaires were gathered in the classroom, using a tablet, in the presence of the class teacher. After explaining the purpose of the research, giving some instructions, guaranteeing anonymity, freedom of participation, and the possibility of interrupting the study at any time, the pupils answered the various questions individually. The presence of a researcher was guaranteed to answer potential questions of understanding.

The self-reported questionnaire consists of four separate parts. Section 1 collects socio-demographic data. Section 2 measures student burnout using the School Burnout Inventory. Section 3 measures school stress using the Daily Worry Scale. Section 4 measures interpersonal variables through questions on support. Finally, the fifth part measures several variables related to the perception of the COVID-19 pandemic.

2.2. Personal Variables

The personal variables were coded on the basis of socio-demographic data (gender, age, nationality, educational achievements) and five questions focusing on summative symptoms (e.g., "I often lack energy; I have stomach, back or head pain; I feel tired; I have difficulty sleeping; I feel fear or anxiety").

2.3. School Burnout

School burnout was measured using the School Burnout Inventory (SBI) by Salmela-Aro, Kiuru, et al. [3] translated and validated in French by Meylan et al. [7]. Cronbach's alpha for the French SBI scale is very satisfactory (0.82), as is Jöreskog's rho (0.91). The SBI comprises 9 items, which assess three dimensions of burnout, emotional exhaustion, cynicism, and feelings of inadequacy. Emotional exhaustion is assessed by four items (e.g., "I feel overwhelmed by my school work"), cynicism by three items (e.g., "I continually wonder if my school work has any meaning"), and feelings of inadequacy by two items (e.g., "I often feel inadequate in my school work"). Each item is scored on a 6-point Likert scale, ranging from 1 to 6 (1 = completely false; 6 = completely true) with a possible total score of 54. The sum of all items provides a total score indicating the severity of the pupils' burnout level: an overall score below 29 represents a low level, a score between 30 and 34 represents an average level, and a score above 34 represents a high level of burnout [7].

2.4. School Stress

School-related stress was measured by the Echelle des tracas quotidiens [38]. Inspired by the Adolescent Hassles Inventory [39] this scale was designed, translated, and validated in French by Plancherel et al. [38]. In the present study, as in Lacombe and Squillaci's study [9], only the 25 items related to the school context were retained, in accordance with the objectives of the study. The items ask about relational difficulties with peers, with teachers or with parents on themes related to school, work, and academic and professional future. The scale measures both the frequency and intensity of daily worries of children and teenagers. For each item, pupils mention whether they have encountered difficulties (or not) in the past 6 months and their rating of how disturbed (hindered) they were. Each item is scored on a 5-point Likert scale from 0-4 (0 = I have not had this problem; 1 = I have had this problem, but it has not bothered me; 4 = I have had this problem and it has bothered me a lot) [38]. In this study, the SBI scale obtained a Cronbach's alpha of 0.89, which represents a strong degree of internal consistency. The sub-dimensions reported a Cronbach's alpha of 0.75 for emotional exhaustion, 0.80 for cynicism, and 0.73 for inadequacy. The daily hassles scale scored 0.89.

2.5. Interpersonal Variables

Interpersonal factors were assessed through questions on perceived support provided from friends, teachers, peers, and parents (e.g., "Do you feel supported by your friends?"). All of these items were scored on a 6-point Likert scale, from 1 to 6 (1 = completely false; 6 = completely true). The perception of the class climate was also collected by the question "do you like your class" (yes/no).

2.6. Organizational Variables

The organizational variables that were collected concerned the number of pupils per class and the number of teachers working in the class during a week.

2.7. Pandemic-Related Variables

Five factors were used to assess students' experiences during the COVID-19 lockdown. Firstly, two questions asked about satisfaction with the supports in place "now" and during the lockdown ("I think the academic supports in place now to catch up on school material are sufficient"; "I think the academic supports in place for organizing distance learning during the lockdown were sufficient"). Secondly, three questions were asked about the perceived return to normality ("I think I have returned to my life as it was before the COVID-19: leisure activities, friendship, socializing"), about the perceived workload ("My workload has increased because of the health crisis"), and about the confidence in the future. All these items were scored on a 6-point Likert scale, from 1 to 6 (1 = completely false; 6 = completely true).

2.8. Statistical Analysis

The significance level was set at 0.05. Bonferroni correction was applied to all post-hoc tests to correct for cumulative error. The effect sizes reported correspond to Cohen's d. To answer the first research question (prevalence of burnout), students were classified according to the degree of burnout they felt (low, medium, high). A score below 29.99 corresponds to a low level of burnout, a score between 30 and 34 corresponds to a medium level, and a score above 34 ($\geq$34.01) corresponds to a high level of burnout [7]. To answer the second research question (comparison of mean scores between T1 and T2), *t*-tests were performed. Finally, for the third research question (risk and protective factors), *t*-tests for mean scores related to categorical variables (such as gender) and correlation analyses (for metric variables) were first performed to analyze single effects and then all significant variables related to personal, interindividual, organizational, and pandemic-related factors were integrated into four separate general linear models in order to combine categorical and metric variables and to analyze the predictive factors of school burnout. Descriptive analyses were performed in Excel and statistical analyses in SPSS 26.

3. Results

The results of the 2nd measurement time (T2) show a proportion of students with severe burnout in the sample of 11.4% (N = 21) out of the total sample (N = 184). Table 2 provides a comparison of the burnout levels of the samples at T1 (2014; N = 313) [9] and T2 (2022; N = 184). The Chi^2 test indicates that the prevalences are significantly different between the two measurement times (x^2 = 11.347 (2) p = 0.003)

Table 2. Prevalence of burnout at T1 and T2.

Level of Burnout	Prevalence T1 (2014) N = 313	Prevalence T2 (2022) N = 184
low	91.4% *	82.6% *
medium	4.8% *	6.0% *
severe	3.8% *	11.4% *
Total	100%	100%

Note. T1 = Measurement time 1; T2 = Measurement time 2; * $p < 0.05$

To compare the mean scores of burnout and its subdimensions over the two measurement times, *t*-tests were performed. Table 3 indicates that the overall burnout score and its subdimensions in 2022 are significantly higher than in 2014. It is however worth noting that this score reflects a low level of burnout in the whole sample according to the cut-off points established by [7].

Table 3. Descriptive data and *t*-test results in the three dimensions of burnout.

	T1 (*N* = 313)		T2 (*N* = 184)			
	M	*SD*	*M*	*SD*	t(...)	*d*
Burnout	1.97	0.85	2.33	1.05	3.929 ***	0.365
EE	1.87	0.84	2.11	1.00	2.692 **	0.25
CYN	1.91	1.08	2.39	1.23	4.465 ***	0.415
INAD	2.28	1.28	2.67	1.43	3.111 **	0.289

Note. ** $p < 0.01$; *** $p < 0.001$; EE = Exhaustion; CYN = Cynicism; INAD = Inadequacy.

Burnout Factors

Within individual factors, gender, somatic symptoms, academic performance, and perceived stress were analyzed in relation to the three dimensions of burnout. The results show that the mean scores between girls and boys do not vary significantly, either in global burnout (Boys M = 2.31; Girls M = 2.34) or in any of the three dimensions of the model.

Then, correlations were conducted to analyze the links between school burnout and the 4 categories of risk factors (individual, interpersonal, organizational, and COVID-19 related). Correlations between burnout and individual risk factors are all statistically significant (Table S1). Regarding the somatic symptoms, results show the more the pupils report symptoms, the more they present high burnout levels. More specifically, lack of energy is more highly related to total burnout and cynicism (521 **). Fatigue is more strongly related to burnout and to emotional exhaustion (0.554 **). Regarding academic performances (in mathematics and in French), the correlations results are all negative and significant. These findings highlight links between high academic performance and low level of burnout. It is to be noted that the effect sizes of the correlations in French performance (−0.273 **) are smaller than those in mathematics (−0.460 **). Finally, correlations between students' overall perceived stress and the level of burnout were analyzed. Several forms of stress have been identified through the Daily Worry Scale. All correlations were significant and positive, suggesting that perceived stress levels are related to burnout and its subdimensions. Success stress ("pressure to perform") was the most strongly correlated with burnout (0.683 **), emotional exhaustion (0.657 **), and feelings of inadequacy (0.639 **). Relational stress was strongly related to emotional exhaustion (0.529 **), underlining the relevance of relationships for the pupils' emotional well-being.

Of the interpersonal factors, support from teachers, classmates, friends, and parents were assessed. All the correlations were found to be statistically significant (Table S1). Correlations are negative, meaning that the higher the perception of support, the lower the levels of burnout, both in total burnout and in the sub-dimensions. The strongest correlations are those of classmates and then of teachers. Finally, the classroom climate, measured with the item "Do you like your classroom" was subjected to a *t*-test. Results show statistically significant differences ($t(182) = -4.194$, $p < 0.001$) in burnout means between pupils who declare liking their class ($M = 2.22$, $SD = 1.00$) and those who do not ($M = 3.22$, $SD = 0.99$). Pupils who declare not appreciating their classroom climate scored significantly higher means for emotional exhaustion ($M = 2.84$, $SD = 1.26$; $t(182) = -3.588$, $p < 0.001$), for cynicism ($M = 3.37$, SD = 1.15; $t(182) = -3.880$, $p < 0.001$) and for inadequacy ($M = 3.75$, $SD = 1.20$; $t(182) = -3.689$, $p < 0.001$) than those who declare that they enjoy it ($M = 2.02$, $SD = 0.93$; $M = 2.28$, $SD = 1.19$; $M = 2.54$, $SD = 1.40$, respectively).

At the level of organizational factors, two variables related to the school context were tested: the number of students per class and the number of teachers per class. However, the analysis of variance showed a non-significant effect of the number of teachers on burnout ($F(3,180) = 0.82$, $p = 0.484$) and its dimensions, namely emotional exhaustion ($F(3,180) = 1.21$, $p = 0.308$), cynicism ($F(3,180) = 0.14$, $p = 0.936$), and inadequacy ($F(3,180) = 1.05$, $p = 0.373$). Regarding the number of students per class, the ANOVA results were also non-significant. The number of students did not influence burnout ($F(3,180) = 1.16$, $p = 0.326$) as well as feelings of emotional exhaustion (F(3,180) = 0.72, p = 0.539), cynicism (F(3,180) = 0.97, p = 0.407) and inadequacy of the students in this sample (F(3,180) = 1.59, p = 0.193). The average tables are available (Table S2).

School burnout means were compared with the five items related to the COVID-19. Findings indicate that perceived increased workload is positively correlated with school burnout (0.402 **). Being confident about the future (−0.551 **) and feeling a sense of having regained their pre-pandemic life (−0.495 **) are the factors most significantly negatively related to burnout. This suggests the more positively future confidence is rated, the more positive pupils report having regained their pre-COVID life, the lower burnout is expressed. Adequacy of current supports was significantly and negatively correlated with burnout, emotional exhaustion, cynicism, and inadequacy.

All the significant variables were fitted to a general linear model to assess the predictive value of the overall pupil burnout (Table 4).

Table 4. Results of the general linear model between the different risk factors and burnout.

	B	*SE*	*t*	*p*	**Eta-Square**
Personal factors					
Somatic symptoms					
Lack of energy	0.048	0.049	990	0.323	-
Pain (stomach, head, back)	0.080	0.033	2.422	0.016	0.033
Fatigue	0.101	0.045	2.227	0.027	0.028
Sleep difficulties	0.067	0.033	2.018	0.045	0.023
Fear, anxiety	0.126	0.048	2.744	0.007	0.042
School results					
Mathematics average	−0.386	0.081	−4.771	<0.001	0.117
French average	0.038	0.080	0.470	0.639	-
Perceived stress					
Total perceived stress	0.684	0.113	6.056	<0.001	0.177
Interpersonal factors					
Support from teachers	−0.135	0.057	−2.345	0.020	0.030
Support from parents	−0.290	0.059	−4.923	<0.001	0.119
Support from friends	−0.110	0.062	−1.767	0.079	-
Support from peers	−0.238	0.056	−4.235	<0.001	0.091
Classroom climate	−0.811	0.219	−3.698	<0.001	0.071
Pandemic factors					
Increased workload since COVID-19	0.174	0.054	3.217	0.002	0.055
Back to life as before COVID-19	−0.231	0.048	−4.779	<0.001	0.114
Adequacy of current support	−0.124	0.055	−2.271	0.024	0.028
Sufficiency of support during containment	−0.013	0.045	−0.282	0.779	-
Trust in the future	−0.378	0.065	−5.830	<0.001	0.160

A general significant model was found for the personal factors (B = 2.363 (SE = 0.432), t = 5.472, p < 0.001). In terms of somatic functioning, pain, fatigue, sleep difficulties, and fear were found to be predictive of school burnout. Among somatic symptoms, fear, and anxiety are the most predictive of burnout, but the effect size remains medium and is small for other somatic symptoms. In terms of school results, poor performance in mathematics predicted student burnout. Effect size is important for performance in mathematics. It is interesting to observe that the French scores are not predictive of the level of burnout here although the correlation was found to be significant. Finally, perceived high stress is the strongest predictor of burnout, which is consistent with the definition of burnout itself, since it is stress that exceeds the individual's resources.

A general significant model emerged between school burnout and interpersonal factors ($B = 6.127$ ($SE = 0.378$), $t = 16.225$, $p < 0.001$). Findings indicate (Table 4) that the perceived supports from teachers, classmates, and parents are negative predictive factors for burnout acting as protective factors. In this sample, parent and classmate support had the greatest impact. Teacher support is also a protective factor, but the effect size is medium in this sample. Finally, as shown in Table 4, results highlight that a positive classroom climate perceived is a negative predictor of burnout, which means that pupils with a positive classroom climate are less likely to experience burnout than their peers

A general significant model emerged between school burnout and COVID-19 related factors ($B = 5.082$ (SE = 0.406), $t = 12.503$, $p < 0.001$). More specifically, findings (Table 4) indicate that feeling that one's daily life is back to normal, the adequacy of existing supports and the confidence about the future were found to be negative predictive factors for burnout, whereas feeling an increased workload caused by the COVID-19 was found to be predictive of school burnout. It is interesting to note that the two most protective factors against burnout are the feeling that one's daily life is back to normal. According to the effect size, this factor would be as strong as the mathematics results. In addition, having confidence in the future appears to be the second most important protective factor of all.

4. Discussion

Burnout is a syndrome that affects not only people in many careers, but also school-aged children. Research has shown that pupils can suffer from burnout most likely due to personal and interpersonal factors. While many studies have aimed at identifying and better understanding the particularly high risk for pupils, the present study is the first to compare the risk of burnout for preteens at two different measurement times and attempt to identify risk factors related also to the consequences of the COVID-19.

4.1. Burnout Prevalence

In the sample, 11.4% of students mentioned high levels of burnout ($M = 12.38$, $SD = 0.48$). In conformity with our hypothesis, the prevalence of burnout has significantly increased between 2015 and 2022. This result is in line with the Swiss national report [40], which found that 88.5% of children aged 11 to 15 considered themselves to be in good psychological health and that the prevalence of young people with depressive symptoms has increased. This result is slightly lower than those in the literature. The main hypothesis is that of age, as studies show an increase in burnout in the adolescent period. Indeed, as noted by Gabola et al. [8] and confirmed by numerous studies [37,41,42]. The older the students are, the more likely they are to experience high levels of burnout.

With regard to the overall burnout score ($M = 20.96$; $SD = 9.43$) measured in the sample, it is worth noting that this score corresponds to a low level of burnout in the whole sample according to the thresholds established by Meylan, Doudin, Antonietti et al. [7]. These data are consistent with the results of the literature which also find low overall burnout scores if the whole sample is taken into consideration [6,9,43–45]. The hypothesis is therefore validated. This low burnout rate in the overall sample can be related to the number of young people who have a positive attitude towards school. The Swiss national report [40] shows that 76% of young people like school. Comparing the overall scores measured in Switzerland among 8th grades pupils in 2014 and those measured in 2022, a significant increase is observed. This may suggest an effect in the mid-term of COVID-19 on the perceived health of preteen students at school. However, other factors could explain this difference. The latest national report on the health of children and adolescents in Switzerland [40] shows that the proportion of young people aged 11 to 25 suffering from psycho-emotional disorders (sadness, nervousness, anxiety, difficulty in falling asleep, depression) has increased over the last 10 years. The report also notes that excessive use of digital media can lead to feelings of loneliness and depression. Therefore, student burnout will be measured again in a similar sample in 2023, 2024, and 2025 as a follow-up to this research.

4.2. Burnout Factors

With regard to individual variables, the results do not show a gender difference. In the literature, only few studies report a higher level of burnout in girls [11,13], the most do not find a significant difference by gender [8,9,12,20,21,42]. Findings of this research are therefore consistent with the trend observed in the literature and validate our hypothesis. In terms of somatic symptoms, lack of energy seems to be related to cynicism, which corroborates the findings of the literature since cynicism is characterized by a loss of interest and motivation for school [23] Fatigue and anxiety are more strongly linked to the feeling of emotional exhaustion, which is precisely characterized by fatigue linked to the perception of being overworked [5]. The perception of being inadequate as a student (inadequacy) is strongly correlated in this study with the feeling of anxiety. In this respect, pain, fatigue, sleep difficulties, and fear appear to be predictive factors of school burnout for the students.

The impact of school performance, especially in mathematics, on burnout is high, since low performance is a predictor of burnout. These results confirm data from the literature and validate our hypothesis. For Virtanen et al. [12], academic performance is positively associated with cognitive and behavioral engagement. According to these authors, students with good results are more engaged, value school more, and use their resources better. Moreover, it is interesting to note that in our sample, inadequacy is the burnout dimension most strongly correlated with academic performance, and this is both in mathematics and in French. As mentioned by Maslach et al. [46], inadequacy is the self-reported dimension of burnout. These results suggest that a low-achieving child appears to perceive him/herself as less academically able and less accomplished than a pupil with high perceived achievement. In this regard, several authors [12,47,48] highlight the protective effect of high self-esteem and a high sense of self-efficacy on burnout. The results also show that school stress is a predictive factor for burnout, which is consistent with findings from the literature that define burnout as "a chronic stress syndrome" [23] (p. 33).

Among the interpersonal variables considered in the study, different sources of support were analyzed to understand their effects on burnout. All sources of support—parents, teachers, friends, peers—were found to be significantly and negatively correlated with burnout and its dimensions. These results validate our hypothesis. It is worth noting that only perceived lack of support from parents, peers, and teachers predicted school burnout. The protective factor with the largest effect size is parental support. This result confirms Sim's [49] finding that the most important social support was that of the parents. For Zakary and Bendahman [50] the more parents take an interest in school work and trust their child, the more successful the pupils are. Conversely, lack of parent supports become a risk when parents devaluate their child or place high pressure on academic performance [51]. The study by Meylan et al. [6] show that parental support negatively predicts burnout and more specifically the level of cynicism and inadequacy. In our study, 85% of children say they are strongly supported by their family, which is close to the 90% reported by the Swiss national report on child and adolescent health [40]. In terms of support received in the school setting, the literature confirms that the more support students perceive from their teachers, the lower their levels of burnout [1]. According to Salmela-Aro et al. [1] teacher motivation, encouragement, and interest is negatively related to students' burnout scores. Conversely, teacher psychological control increases student emotional exhaustion and cynicism [52]. In the research by Meylan et al. [6], the most important effect of perceived lack of support also concerns cynicism. It can be hypothesized that when support is perceived negatively, students become disengaged and feel more cynical towards school. In our study, 69.6%

of students report a high degree of support from their teacher. This result can be put into perspective with the Swiss national report of 2020 [40]. In the latter, the majority of 11–13-year-olds (70%) say they have confidence in their teachers, while at 14–15 years old 57% of boys and 52% of girls have this opinion. This result may also explain the increase in burnout among adolescents compared to 11–13-year-olds.

The results indicate that peer support appears to be a protective factor against burnout in all three dimensions of the model. According to Estell and Perdue [53], peer support is thought to foster attachment and engagement in school, feelings that protect against burnout and its consequent symptoms. The results of the literature also point to the importance of support from peers who are positively engaged with school [12].

Finally, our results indicate that the assessment of the classroom climate influences the burnout averages, both in terms of emotional exhaustion, cynicism and inadequacy. These results corroborate those of the literature in particular the results obtained in our previous study conducted in 2014 [9] and tend to confirm our hypothesis. While a positively perceived classroom climate seems to protect students from burnout, a negative atmosphere can have the opposite effects [1] such as rumors spreading, a recognized risk factor in research [54].

With regard to organizational factors, the two factors investigated in this study were the number of pupils per class and the number of teachers per week. The results indicate that class size does not significantly affect the means of burnout, emotional exhaustion, cynicism, and inadequacy. Although class size reduction can improve academic performance [55], no study seems to have linked this factor to the problem of school burnout. As for the number of teachers worked with per week, this variable also does not lead to a significant difference between the averages for burnout, emotional exhaustion, cynicism and inadequacy. These results suggest that school burnout and its constituent symptoms are not influenced by quantity, but rather by the quality of interpersonal relationships.

Finally, the variables related to the COVID-19 pandemic show significant correlations with burnout, allowing the identification of one risk factor (workload) and four protective factors (getting back to life before COVID-19; adequacy of current support; adequacy of support during containment; trust in future). In view of the significant increase in the prevalence of burnout among students, these factors provide some insights into how the COVID-19 pandemic appears to have impacted the perceived health in the sample. The feeling of increased workload due to the COVID-19 was found to be a predictor of student burnout. These results are consistent with those of the study of Meylan and Hascoët [56] who report a more specific effect on emotional exhaustion. Indeed, according to Teuber et al. [57], Núnez-Regueiro [58], and Demerouti et al. [43], a high perception of demands is a risk factor for emotional exhaustion and student well-being. Meylan and Hascoët [56] find, among other variables, that workload is related to stress, with stress itself being associated with burnout and its three dimensions.

In response to perceived workload, the adequacy of faculty supports was assessed and the effect on burnout analyzed. Sufficiency of perceived current supports was found to be a negative predictor of burnout. As noted by Salmela-Aro and Upadyaya [59], the perception of sufficient and adequate resources enables the student to engage academically and manage perceived demands effectively. Finally, confidence in the future is a negative predictor of burnout. This variable therefore seems to act as a protective factor against burnout. These results corroborate those of Martos Martínez et al. [47], according to which future prospects and burnout are negatively correlated. According to these authors, students who perceive the future less positively experience more exhaustion, cynicism, and inadequacy than those who perceive it positively.

5. Conclusions

This research measured student burnout two years after the pandemic lockdown and identify risk and protective factors that could explain this syndrome. "The real measure of a country is the attention it gives to its children, their health and safety, their material, their health and safety, their material situation, their education and socialization, and their sense of being loved, valued and to be loved, valued and included in the families and societies into which they are born" [60] (p. 1). This quote alone shows the importance of regularly measuring school burnout in the same population in order to understand its evolution and to propose effective interventions. Results show a significantly higher average burnout than that measured in 2014 among the students surveyed, suggesting a possible effect of the COVID-19 period on students' well-being in the medium term. Several factors were found to be predictive of burnout. At the individual level, perceived stress, low academic performance in mathematics and somatic symptoms such as lack of energy, fatigue, pain, sleep difficulties, and anxiety predict the onset of school burnout. At the interpersonal level, support from classmates, teachers, and parents are related to a decrease in burnout and summative symptoms and a negative classroom climate predicts student burnout. At the organizational level, no factor was found to be significant. Finally, with regard to factors related to the COVID-19 pandemic, students who indicated that they had returned to their pre-COVID-19 life, had confidence in the future, and received sufficient support to make up for their shortcomings were less likely to suffer burnout than their peers. These three factors are negative predictors of burnout. Conversely, an increase in workload due to COVID-19 is a predictor of burnout.

At the end of this research, several limitations should be mentioned. The first limitation is the type of questionnaire used. Indeed, these are self-report tools filled in by the students about themselves. There is therefore a risk of subjectivity in the answers. In addition, the variables measured in relation to the COVID-19 pandemic do not come from a standardized questionnaire, so further research would be necessary. Notwithstanding these limitations, the strength of the study is in having measured the means of two samples with comparable characteristics (age, education level, district, school environment, etc.), in order to provide a comparison on the burnout of two group of students. Another strength was to compare these scores once the pandemic was over, in order to analyze whether perceived health had returned to pre-pandemic normality two years later and, if not, to identify factors related to the syndrome. Finally, our study also points to some factors that would be interesting to study in the future, such as the level of teacher burnout in which the pupils are, or the impact of the use of digital media at home, which have been shown to be related to sleep difficulties and feelings of loneliness in previous studies [40].

Educational Implications

Our results highlight some important points. Firstly, school burnout should not be neglected. It already appears in young students [9,16,61] and tends to increase over the years of study [41,48]. Thus, it is important to study young populations, in order to continue the knowledge of the phenomenon and specially to identify the first signs of emotional exhaustion, cynicism, and inadequacy in these students so as to be able to act as soon as the first symptoms appear to limit their consequences. Secondly, it is important to regularly measure the prevalence and scores of burnout in similar samples. Indeed, if this study has highlighted an increase in school burnout among preteen students between 2014 and 2022, repeated measurements make it possible to monitor the evolution of mental health reported by students, especially during periods of health, economic, or social crisis. Thirdly, this study highlights the large number of explanatory factors of school burnout. While the study of risk factors is important, the study of protective factors is equally important, particularly in post-crisis periods [42], so that supports can be adjusted by putting in place services that meet the needs of students in order to promote their development even in crisis situations [62]. This study also opens up perspectives for intervention. As mentioned, the factors with the greatest impact are: confidence in the future, perceived stress, parental support, and mathematics results. It is therefore possible to imagine an intervention targeting these four factors or specifically one of them. For example, it would be interesting to involve parents in schooling by offering open classes where parents could see their child in the school setting. Another possibility would be an intervention targeting mathematical ability. In this regard, many literature reviews highlight evidence-based interventions to improve mathematics achievement [63–65]. In terms of stress, it would be possible to teach coping strategies and to regularly measure students' perceived stress and the causes of it. This would make it possible to intervene directly before the onset of school burnout. Finally, in terms of confidence in the future, it would be interesting to ask students about their desires for the future in order to set up projects that would motivate them (for example, by proposing internships in different professions).

Supplementary Materials: The following supporting information can be downloaded at: https://www.mdpi.com/article/10.3390/children10050823/s1, Table S1: Results of correlation between burnout and risk factors; Table S2: Results of ANOVA between burnout and organizational factors.

Author Contributions: Conceptualization, N.L. and M.S.; methodology, N.L.; formal analysis, V.H.; investigation, M.H.; data curation, N.L. and M.H.; writing—original draft, N.L.; writing—review and editing, C.P. and M.S.; supervision, M.S.; project administration, M.S. All authors have read and agreed to the published version of the manuscript.

Funding: This research received no external funding.

Institutional Review Board Statement: Ethical review and approval were waived for this study due to the fact that no personal data were collected (neither name, nor surname, nor photos, nor videos). It is therefore not possible to identify the participants.

Informed Consent Statement: Informed consent was obtained from all subjects involved in the study.

Data Availability Statement: The data presented in this study are available on request from the corresponding author. The data are not publicly available due to the fact that they are children and that the research indicated that the data would only be stored on a secure server at the university of Fribourg.

Conflicts of Interest: The authors declare no conflict of interest.

References

1. Salmela-Aro, K.; Kiuru, N.; Pietikäinen, M.; Jokela, J. Does School Matter? The Role of School context in Adolescents' school-related burnout. *Eur. Psychol.* **2008**, *13*, 12–23. [CrossRef]
2. Parker, P.D.; Salmela-Aro, K. Developmental processes in school burnout: A comparison of major developmental models. *Learn. Individ. Differ.* **2011**, *21*, 244–248. [CrossRef]
3. Salmela-Aro, K.; Kiuru, N.; Leskinen, E.; Nurmi, J.E. School Burnout Inventory (SBI): Reliability and Validity. *Eur. J. Psychol. Assess.* **2009**, *25*, 48–57. [CrossRef]
4. Maslach, C.; Jackson, S.E. The measurement of experienced burnout. *J. Organ. Behav.* **1981**, *2*, 99–113. [CrossRef]
5. Lebert-Charron, A.; Boujut, E.; Beaudoin, M.; Quintard, B. Du stress au burnout scolaire: Définitions, déterminants pychosociaux, évaluation et pistes de prévention. *Prat. Psychol.* **2021**, *27*, 57–69. [CrossRef]
6. Meylan, N.; Doudin, P.-A.; Curchod-Ruedi, D.; Stephan, P. Burnout scolaire et soutien social: L'importance du soutien des parents et des enseignants. *Psychol. Française* **2014**, *60*, 1–15. [CrossRef]
7. Meylan, N.; Doudin, P.-A.; Antonietti, J.-P.; Stéphan, P. Validation française du School Burnout Inventory. *Eur. Rev. Appl. Psychol.* **2015**, *65*, 285–293. [CrossRef]
8. Gabola, P.; Meylan, N.; Hascoët, M.; De Stasio, S.; Fiorilli, C. Adolescents' School Burnout: A Comparative Study between Italy and Switzerland. *Eur. J. Investig. Health Psychol. Educ.* **2021**, *11*, 849–859. [CrossRef]
9. Lacombe, N.; Squillaci, M. Liens entre un examen de passage (PPO) et le burnout scolaire. *Rev. Suisse Pédagogie Spécialisée* **2015**, *4*, 15–21. Available online: https://edudoc.ch/record/119852/ (accessed on 20 January 2022).
10. Salmela-Aro, K.; Tynkkynen, L. Gendered pathways in school burnout among adolescents. *J. Adolesc.* **2012**, *35*, 929–939. [CrossRef]
11. Kinnunen, J.M.; Lindfors, P.; Rimpelä, A.; Salmela-Aro, K.; Rathmann, K.; Perelman, J.; Federico, B.; Richter, M.; Kunst, A.E.; Lorant, V. Academic well-being and smoking among 14- to 17-year-old schoolchildren in six European cities. *J. Adolesc.* **2016**, *50*, 56–64. [CrossRef] [PubMed]
12. Virtanen, T.E.; Lerkkanen, M.-K.; Poikkeus, A.-M.; Kuorelahti, M. Student Engagement and School Burnout in Finnish Lower-Secondary Schools: Latent Profile Analysis. *Scand. J. Educ. Res.* **2016**, *62*, 519–537. [CrossRef]
13. Gerber, M.; Colledge, F.; Mücke, M.; Schilling, R.; Brand, S.; Ludyga, S. Psychometric properties of the Shirom-Melamed Burnout Measure (SMBM) among adolescents: Results from three cross-sectional studies. *BMC Psychiatry* **2018**, *18*, 1–13. [CrossRef] [PubMed]
14. Lee, M.Y.; Lee, M.K.; Lee, M.J.; Lee, S.M. Academic Burnout Profiles and Motivation Styles among Korean High School Students. *Jpn. Psychol. Res.* **2019**, *62*, 184–195. [CrossRef]

15. Cheung, P.; Li, C. Physical Activity and Mental Toughness as Antecedents of Academic Burnout among School Students: A Latent Profile Approach. *Int. J. Environ. Res. Public Health* **2019**, *16*, 2024. [CrossRef]
16. Yildiz, V.A.; Kiliç, D. Investigation of School Burnout and School Attachment among Secondary School Students. *Educ. Sci. Theory Pract.* **2020**, *20*, 44–55. [CrossRef]
17. Dompierre, J.; Lavoie, F.; Pérusse, M. Les déterminants individuels, interpersonnels et organisationnels de la détresse psychologique en milieu de travail. *Can. Psychol. Can.* **1993**, *34*, 365–380. [CrossRef]
18. Schaufeli, W.B. Past Performance and Future Perspectives of Burnout Research. *SA J. Ind. Psychol.* **2003**, *29*, 1–15. [CrossRef]
19. Slivar, B. The syndrome of burnout, self-image, and anxiety with grammar school students. *Horiz. Psychol.* **2001**, *10*, 21–32. Available online: http://psiholoska-obzorja.si/arhiv_clanki/2001_2/slivar.pdf (accessed on 20 January 2022).
20. Meylan, N.; Doudin, P.-A.; Curchod, D.; Stephan, P. School burnout in adolescents: Differences in background variables and exploration of school-related stress at the end of compulsory schooling. *Ric. Psicol.* **2011**, *4*, 539–563. [CrossRef]
21. Zakari, S.; Walburg, V.; Chabrol, H. Study of school burnout, depression and suicidal ideation in French high school students. *J. De Thérapie Comport. Et Cogn.* **2008**, *18*, 113–118. [CrossRef]
22. Kiuru, N.; Leskinen, E.; Nurmi, J.-E.; Salmela-Aro, K. Depressive symptoms during adolescence: Do learning difficulties matter? *Int. J. Behav. Dev.* **2011**, *35*, 298–306. [CrossRef]
23. Salmela-Aro, K. Le burnout scolaire. In *La Santé Psychosociale des Elèves*; Curchod-Ruedi, D., Doudin, P.-A., Lafortune, L., Lafranchise, N., Eds.; Presses de l'Université du Québec: Montreal, QC, Canada, 2011; pp. 32–45.
24. Kiuru, N.; Aunola, K.; Nurmi, J.-E.; Leskinen, E.; Salmela-Aro, K. Peer Group Influence and Selection in Adolescents' School Burnout: A Longitudinal Study. *Merrill-Palmer Q.* **2008**, *54*, 23–55. [CrossRef]
25. Schaufeli, W.B.; Martínez, I.M.; Pinto, A.M.; Salanova, M.; Bakker, A.B. Burnout and Engagement in University Students: A Cross-National Study. *J. Cross-Cult. Psychol.* **2002**, *33*, 464–481. [CrossRef]
26. Roorda, D.L.; Jak, S.; Zee, M.; Oort, F.J.; Koomen, H.M.Y. Affective Teacher–Student Relationships and Students' Engagement and Achievement: A Meta-Analytic Update and Test of the Mediating Role of Engagement. *Sch. Psychol. Rev.* **2017**, *46*, 239–261. [CrossRef]
27. Meylan, N.; Doudin, P.-A.; Curchod-Ruedi, D.; Antonietti, J.-P.; Stephan, P. Stress scolaire, soutien social et burnout à l'adolescence: Quelles relations? *Éducation Francoph.* **2015**, *43*, 135–153. [CrossRef]
28. Leung, G.S.M.; Yeung, K.C.; Wong, D.F.K. Academic Stressors and Anxiety in Children: The Role of Paternal Support. *J. Child Fam. Stud.* **2010**, *19*, 90–100. [CrossRef]
29. Bask, M.; Salmela-Aro, K. Burned out to drop out: Exploring the relationship between school burnout and school dropout. *Eur. J. Psychol. Educ.* **2013**, *28*, 511–528. [CrossRef]
30. UNICEF. The State of the World's Children 2021. In *My Mind: Promoting, Protecting and Caring for Children's Mental Health*; UNICEF: New York, NY, USA, 2022.
31. UNICEF Innocenti Research Centre. *Life in Lockdown: Child and Adolescent Mental Health and Well-Being in the Time of COVID-19*; United Nations: New York, NY, USA, 2021. [CrossRef]
32. Pro Juventute. Pro Juventute Rapport de Pro Juventute sur le coronavirus: Impacts de la pandémie de Covid-19 sur les enfants, les jeunes et leurs familles en Suisse. 2021. Available online: https://www.projuventute.ch/sites/default/files/2021-02/Pro-Juventute-Corona-Report-FR.pdf (accessed on 7 February 2022).
33. UN. Note de synthèse: L'impact de la COVID-19 sur les Enfants. 15 April 2020. Available online: https://www.un.org/sites/un2.un.org/files/note_de_synthese_-_limpact_de_la_covid-19_sur_les_enfants_0.pdf (accessed on 7 February 2022).
34. Ma, L.; Mazidi, M.; Li, K.; Li, Y.; Chen, S.; Kirwan, R.; Zhou, H.; Yan, N.; Rahman, A.; Wang, W.; et al. Prevalence of mental health problems among children and adolescents during the COVID-19 pandemic: A systematic review and meta-analysis. *J. Affect. Disord.* **2021**, *293*, 78–89. [CrossRef]
35. Tang, S.; Xiang, M.; Cheung, T.; Xiang, Y.-T. Mental health and its correlates among children and adolescents during COVID-19 school closure: The importance of parent-child discussion. *J. Affect. Disord.* **2021**, *279*, 353–360. [CrossRef]
36. Racine, N.; McArthur, B.A.; Cooke, J.E.; Eirich, R.; Zhu, J.; Madigan, S. Global Prevalence of Depressive and Anxiety Symptoms in Children and Adolescents during COVID-19. *JAMA Pediatr.* **2021**, *175*, 1142–1150. [CrossRef] [PubMed]
37. Fiorilli, C.; De Stasio, S.; Di Chiacchio, C.; Pepe, A.; Salmela-Aro, K. School burnout, depressive symptoms and engagement: Their combined effect on student achievement. *Int. J. Educ. Res.* **2017**, *84*, 1–12. [CrossRef]
38. Plancherel, B.; Bettschart, W.; Bolognini, M.; Dumont, M.; Halfon, O. Influence comparée des événements existentiels et des tracas quotidiens sur la santé psychique à la préadolescence. *Child Adolesc. Neuropsychiatry* **1997**, *45*, 126–138.
39. Bobo, J.K.; Gilchrist, L.D.; Elmer, J.F.; Snow, W.H.; Schinke, S.P. Hassles, role strain, and peer relations in young adolescents. *J. Early Adolesc.* **1986**, *6*, 339–352. [CrossRef]
40. Observatoire Suisse de la santé (Ed.) *La Santé en Suisse-Enfants, Adolescents et Jeunes Adultes*; Rapport national sur la santé; Hogrefe: Bern, Swizerland, 2020.
41. Lee, M.Y.; Lee, S.M. The effects of psychological maladjustments on predicting developmental trajectories of academic burnout. *Sch. Psychol. Int.* **2018**, *39*, 217–233. [CrossRef]
42. Ying, L.; Wang, Y.; Lin, C.; Chen, C. Trait resilience moderated the relationships between PTG and adolescent academic burnout in a post-disaster context. *Pers. Individ. Differ.* **2016**, *90*, 108–112. [CrossRef]

43. Demerouti, E.; Bakker, A.B.; Nachreiner, F.; Schaufeli, W.B. The job demands-resources model of burnout. *J. Appl. Psychol.* **2001**, *86*, 499–512. [CrossRef]
44. Fiorilli, C.; Farina, E.; Buonomo, I.; Costa, S.; Romano, L.; Larcan, R.; Petrides, K.V. Trait Emotional Intelligence and School Burnout: The Mediating Role of Resilience and Academic Anxiety in High School. *Int. J. Environ. Res. Public Health* **2020**, *17*, 3058. [CrossRef]
45. Elliot, C.; Lang, C.; Brand, S.; Holsboer-Trachsler, E.; Pühse, U.; Gerber, M. The Relationship Between Meeting Vigorous Physical Activity Recommendations and Burnout Symptoms among Adolescents: An Exploratory Study with Vocational Students. *J. Sport Exerc. Psychol.* **2015**, *37*, 180–192. [CrossRef]
46. Maslach, C.; Schaufeli, W.B.; Leiter, M.P. Job Burnout. *Annu. Rev. Psychol.* **2001**, *52*, 397–422. [CrossRef]
47. Martos Martínez, Á.; Molero Jurado, M.D.M.; Pérez-Fuentes, M.D.C.; Barragán Martín, A.B.; Simón Márquez, M.D.M.; Tortosa Martínez, B.M.; Sisto, M.; Gázquez Linares, J.J. Self-Efficacy, Positive Future Outlook and School Burnout in Spanish Adolescents. *Sustainability* **2021**, *13*, 4575. [CrossRef]
48. Parviainen, M.; Aunola, K.; Torppa, M.; Lerkkanen, M.-K.; Poikkeus, A.-M.; Vasalampi, K. Early Antecedents of School Burnout in Upper Secondary Education: A Five-year Longitudinal Study. *J. Youth Adolesc.* **2021**, *50*, 231–245. [CrossRef] [PubMed]
49. Sim, H.-O. Relationship of Daily Hassles and Social Support to Depression and Antisocial Behavior among Early Adolescents. *J. Youth Adolesc.* **2000**, *29*, 647–659. [CrossRef]
50. Zakari, S.; Bendahman, H. Syndrôme d'épuisement scolaire et dépression à l'adolescence: Étude qualitative auprès d'une vingtaine de lycéens français. *Les Collect. Du Cirp* **2011**, *2*, 318–329.
51. Ang, R.P.; Huan, V.S. Relationship between academic stress and suicidal ideation: Testing for depression as a mediator using multiple regression. *Child Psychiatry Hum. Dev.* **2006**, *37*, 133–143. [CrossRef] [PubMed]
52. Shih, S.-S. The effects of autonomy support versus psychological control and work engagement versus academic burnout on adolescents' use of avoidance strategies. *Sch. Psychol. Int.* **2013**, *34*, 330–347. [CrossRef]
53. Estell, D.B.; Perdue, N.H. Social Support and Behavioral and Affective School Engagement: The Effects of Peers, Parents, and Teachers. *Psychol. Sch.* **2013**, *50*, 325–339. [CrossRef]
54. Wang, Q.; Hu, W.; Ouyang, X.; Chen, H.; Qi, Y.; Jiang, Y. The relationship between negative school gossip and suicide intention in Chinese junior high school students: The mediating effect of academic burnout and gender difference. *Child. Youth Serv. Rev.* **2020**, *117*, 105272. [CrossRef]
55. Terrisse, B.; Boufrahi, S.; Larivée, S.; Lefebvre, M.-L. La réduction des effectifs dans la classe pour favoriser la réussite scolaire: Etude de l'implantation et des retombées de la diminution du nombre d'élèves par classe au préscolaire et au premier cycle du primaire au Québec. In *Rapport Final de Recherche Présenté au Fonds Québécois de la Recherche sur la Société et la Culture (FQRSC) Dans le Cadre du Programme des Actions Concertées*; Université du Québec à Montréal: Montreal, QC, Canada, 2011. Available online: https://core.ac.uk/download/pdf/46923942.pdf (accessed on 7 February 2022).
56. Meylan, N.; Hascoët, M. Stress, burnout and health crisis in the school context: Point of view of students and teachers in French-speaking Switzerland. *Rev. Suisse Des Sci. L'éducation, under press.*
57. Teuber, Z.; Nussbeck, F.W.; Wild, E. School burnout among Chinese high school students: The role of teacher-student relationships and personal resources. *Educ. Psychol.* **2021**, *41*, 985–1002. [CrossRef]
58. Núñez-Regueiro, F. Le décrochage scolaire comme processus de stress: Des profils hétérogènes sous forme d'ennui et de surmenage. *Sch. Career Guid.* **2017**, *46*, 1–23. [CrossRef]
59. Salmela-Aro, K.; Upadyaya, K. School burnout and engagement in the context of demands-resources model. *Br. J. Educ. Psychol.* **2014**, *84*, 137–151. [CrossRef] [PubMed]
60. UNICEF. La Pauvreté des Enfants en Perspective: Vue D'ensemble du Bien-Etre des Enfants Dans les Pays Riches. Bilan Inno-centi 7, UNICEF, Centre de Recherche Innocenti, Florence. 2007. Available online: https://www.unicef-irc.org/publications/pdf/rc7_fre.pdf (accessed on 20 March 2023).
61. Salmela-Aro, K.; Muotka, J.; Alho, K.; Hakkarainen, K.; Lonka, K. School burnout and engagement profiles among digital natives in Finland: A person-oriented approach. *Eur. J. Dev. Psychol.* **2016**, *13*, 704–718. [CrossRef]
62. Stoecklin, D.; Richner, L. Enquête exploratoire—Le vécu des enfants et adolescents de 11 à 17 ans en Suisse romande par rapport au COVID-19 et aux mesures associées (semi-confinement). Université de Genève. Available online: https://www.unige.ch/cide/index.php?cID=583 (accessed on 7 February 2022).
63. Lacombe, N.; De Chambrier, A.-F.; Dias, T. Des données probantes au service de l'enseignement différencié des mathématiques. *Rev. Mathématique Pour L'école* **2021**, *236*, 13–26. [CrossRef]
64. Baker, S.; Gersten, R.; Lee, D.-S. A synthesis of empirical research on teaching mathematics to low-achieving students. *Elem. Sch. J.* **2002**, *103*, 51–73. [CrossRef]
65. Gersten, R.; Chard, D.J.; Jayanthi, M.; Baker, S.K.; Morphy, P.; Flojo, J. Mathematics Instruction for Students with Learning Disabilities: A Meta-Analysis of Instructional Components. *Rev. Educ. Res.* **2009**, *79*, 1202–1242. [CrossRef]

children

MDPI

Article

Exploring the Prevalence and Predictors of Anxiety among Lithuanian Adolescents during Times of Crisis: A Cross-Sectional Study

Laura Šalčiūnaitė-Nikonovė [1,2,*], Monika Žemaitaitytė [1] and Kastytis Šmigelskas [1,2]

1 Health Research Institute, Lithuanian University of Health Sciences, LT-47181 Kaunas, Lithuania
2 Department of Health Psychology, Lithuanian University of Health Sciences, LT-47181 Kaunas, Lithuania
* Correspondence: laura.salciunaite@lsmu.lt

Abstract: Recent events in Europe, like the COVID-19 pandemic and the 2022 Russo–Ukrainian War (RUW), might have sparked anxiety among adolescents. This study aimed to compare anxiety levels in Lithuanian adolescents post-COVID-19 peak (Study 1, October 2021) and during the onset of the RUW (Study 2, April–June 2022). Data from 459 participants in Study 1 and 6637 in Study 2, aged 11–17, were collected through HBSC pilot and national surveys in Lithuania. Self-reported questionnaires assessed anxiety, social media use, stress, loneliness, self-efficacy, and peer support factors. Analysis was conducted using multivariable logistic regressions. Notably, anxiety prevalence in Lithuanian adolescents showed no significant difference between Study 1 and Study 2, stabilizing around 24%. In 2021, stress (OR = 5.89, 95% CI 3.11–11.17), problematic social media use (OR = 4.58, 95% CI 1.89–10.58), and female gender (OR = 2.87, 95% CI 1.58–5.22) significantly predicted anxiety. By 2022, stress (OR = 3.68, 95% CI 3.14–4.30), loneliness (OR = 2.85, 95% CI 2.43–3.35), and lower self-efficacy (OR = 1.40, 95% CI 1.20–1.60) emerged as important predictors. This study enhances our understanding of adolescent anxiety during crises, emphasizing the urgency of addressing multiple factors to manage and support vulnerable youth.

Keywords: anxiety; adolescent; COVID-19; RUW; war; HBSC

Citation: Šalčiūnaitė-Nikonovė, L.; Žemaitaitytė, M.; Šmigelskas, K. Exploring the Prevalence and Predictors of Anxiety among Lithuanian Adolescents during Times of Crisis: A Cross-Sectional Study. *Children* **2024**, *11*, 32. https://doi.org/10.3390/children11010032

Academic Editor: Matteo Alessio Chiappedi

Received: 30 November 2023
Revised: 22 December 2023
Accepted: 26 December 2023
Published: 27 December 2023

1. Introduction

In recent years, Europe has faced a persistent crisis due to unprecedented events, notably the onset of the global COVID-19 pandemic in early 2020 followed by the 2022 Russo–Ukrainian War. Stressful events like these can increase the levels of psychological distress in the general population [1–3], with young people being the significant risk group that requires special attention [2].

Anxiety emerges as a prominent and severe affective symptom among children and adolescents during global crises. While anxiety is a natural response to life's stressors, events like nearby wars or pandemics can exacerbate it into prolonged or maladaptive states. Over the past decade, studies suggest an increasing trend in the prevalence rates of anxiety disorders in adolescents, varying from 6% [4] to 32% [5]. Anxiety disorders are recognized as one of the most disabling conditions in adolescents [6], causing challenges across different psychosocial domains, including academic and social functioning [7] and overall quality of life [8]. Furthermore, elevated anxiety during adolescence can predict future depression, substance abuse, and adult anxiety disorders [9]. Hence, during crises, it becomes crucial to closely observe anxiety symptoms among young individuals.

The profound impact of the COVID-19 pandemic on adolescent anxiety levels is evident, with reported symptomatology rates ranging from 23% to 38% [10,11], indicating a significant rise in mental health concerns [12]. Various pandemic-related factors, such as enforced social restrictions and extended periods at home, negatively affected adolescents' overall well-being [13], intensifying feelings of isolation and anxiety [14,15]. School closures disrupted established routines, contributing to an increase in depression cases [13,16]. Parental distress during prolonged homestays emerged as a significant predictor of heightened anxiety in children [11,17]. Moreover, the surge in social media use, up to 10 h daily for some adolescents during the lockdown, was found to be linked to increased anxiety, disrupted sleep patterns [18–20], and heightened depression prevalence, especially among girls [19]. Finally, misinformation on social media notably added to heightened anxiety, particularly for those struggling to verify information accurately [13,18,21].

The Russo–Ukrainian War, commencing in February 2022, may have further impacted teenagers' mental health, particularly anxiety. While Ukrainians, especially children, directly experience the war's impact, its ongoing nature could heighten anxiety across Europe, notably affecting neighboring regions, as seen in similar events, like the Chornobyl disaster [22–24]. Lithuania's historical experiences with Russian occupation in the 20th century may heighten susceptibility to similar situations, fostering increased feelings of threat and insecurity.

Although the scientific literature examining the direct impact of the Russo–Ukrainian War on adolescent populations remains scarce, there have been studies analyzing university students, providing valuable insights. Reports from Czech Republic universities indicated a significant rise in war-related anxiety levels among students [25,26], with older age and female gender correlating with higher anxiety rates [25,27]. Proximity to conflict zones, as observed in Slovakian students compared to their Czech counterparts, seems to exacerbate these concerns [26]. Social media, akin to its role during the pandemic, notably magnifies anxiety. Platforms like Instagram and TikTok may contribute to heightened anxiety from war-related exposure, echoing past instances where media exposure contributed to increased anxiety [28]. Indeed, recent studies highlighted a correlation between media coverage of the RUW and peritraumatic dissociative experiences and anxiety in adolescents [27,29].

Given the importance of recent events on adolescents' well-being and the current scarcity of research, this study seeks to assess anxiety levels among Lithuanian adolescents following the peak of the COVID-19 pandemic (October 2021) and the initiation of the conflict in Ukraine (April–June 2022). Additionally, this research aims to establish potential mental health and behavioral factors associated with anxiety outcomes, with a particular focus on variables such as excessive social media engagement, feelings of loneliness, stress, self-efficacy, and other putative factors. Finally, this study seeks to determine whether discernible differences in anxiety levels exist between cohorts or whether distinct traits are prevalent within each group. By exploring these differences, it intends to highlight the specific impact of varied events on anxiety and ascertain the potential existence of shared contributing factors across these unique circumstances.

2. Materials and Methods

2.1. Study Process and Sample

Data for this study were collected from Lithuanian schools at two different time points: during the HBSC national pilot survey in December 2021 (Study 1) and during the HBSC main national survey in Lithuania conducted between April and June 2022 (Study 2).

Study 1 was conducted as a pilot project for an upcoming 2022 HBSC study in Lithuania. Seven schools were randomly selected from the national schools' list (two from the largest city, two from other cities, and three from towns) by choosing the first five schools that agreed to participate in the study (school response rate 71.4%). Of the 688 students invited, 459 participated in the survey (student response rate 66.7%). A consort diagram of the study sample is presented in Figure 1.

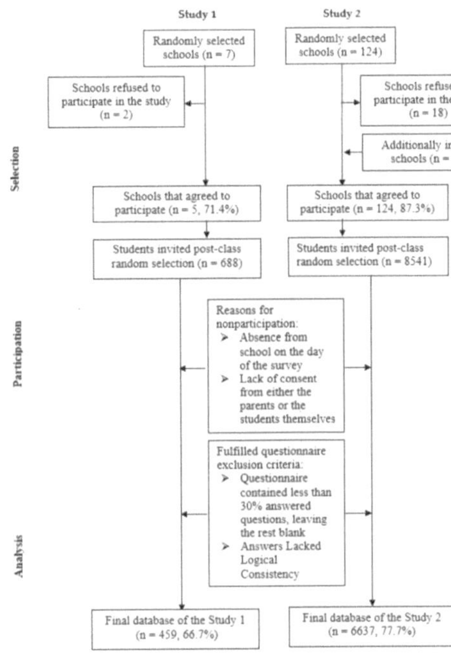

Figure 1. Consort diagram illustrating sample selection for Study 1 and Study 2.

During the main national survey (Study 2) performed to compose the geographically representative sample, a total list of general education and vocational schools in Lithuania was stratified by 15 regions (5 large city municipalities and 10 counties, except for large cities), and then 142 schools were randomly selected for participation in this study. Of them, 124 agreed to participate (school response rate 87.3%). The eligible sample in the agreed-to-participate schools was 8541 students, but some students did not attend school on the day of the survey, refused to participate, or returned corrupt questionnaires, resulting in 6637 students being included in the analysis (student response rate 77.7%) (Figure 1).

In accordance with the HBSC study protocol, the questionnaire was administered in every school to students in the 5th, 7th, 9th, and 11th grades, corresponding to the predominant ages of 11, 13, 15, and 17 years, respectively [30]. The 2021 and 2022 samples followed the same protocol requirements, so the distribution of study participants by gender and class was similar.

2.2. Measures

Paper-based self-report questionnaires were used following the internal protocol recommendations of the HBSC study [30]. For the current analysis, the selected indicators were anxiety, problematic social media use, stress, loneliness, self-efficacy, peer support, and sociodemographic indicators such as gender, grade, and family socioeconomic status.

Anxiety was assessed using the seven-item Generalized Anxiety Scale 7 (GAD-7) [31]. Study participants were asked to rate seven anxiety symptoms over the past two weeks on a 4-point scale ranging from 0 'not at all' to 3 'almost every day'. Responses were summed and grouped into two categories: (1) no–mild anxiety (0–10 points) and (2) moderate–severe anxiety (11–21 points) [30,32]. The internal consistency of the scale in Study 1 was $\alpha = 0.865$, and in Study 2 it was $\alpha = 0.889$.

Problematic social media use was assessed using the nine-item Social Media Disorder (SMD) scale with dichotomous (no/yes) response options [33]. Research participants who answered positively to at least 6 statements were coded as problematic social media users

(PSMUs). The internal consistency of the scale in Study 1 was $\alpha = 0.813$, and in Study 2 it was $\alpha = 0.770$.

Stress was assessed using the four-item Cohen Perceived Stress Scale (PSS-4) [34]. Respondents rated how often they experienced stressful situations on a 5-point Likert scale ranging from 'never' to 'very often' in the last month. The PSS-4 is used as a continuous scale (sum score after reverse-coding the 2nd and 3rd items), where higher scores reflect higher levels of perceived stress. In this study, responses were summed and grouped into two categories: (1) low-level stress (0–8 points) and (2) high-level stress (9–16 points). The internal consistency of the scale in Study 1 was $\alpha = 0.566$, and in Study 2 it was $\alpha = 0.590$.

Loneliness was measured by asking students how often they felt lonely in the past 12 months using a 5-point scale ranging from 'never' to 'always' [35]. The answers of the subjects were divided into 2 groups: 'high level', those who indicated that they always or mostly feel lonely, and 'low', those who indicated that they never, rarely, or sometimes feel lonely.

Self-efficacy was measured using two questions about problem solving and decision implementation. Study participants were asked to rate their abilities on a 5-point scale ranging from 'never' to 'always'. Their responses were divided into 2 groups: 'high', those who indicated that they always or most of the time managed things successfully, and 'low', those who indicated that they never, rarely, or sometimes managed things successfully [30].

Peer support was assessed using one subscale from the Multidimensional Scale of Perceived Social Support (MSPSS) [36]. Study participants were asked to rate their answers on a 7-point Likert scale ranging from 'very strongly disagree' to 'very strongly agree'. High peer support was defined as a mean score of 5.5, indicating high peer support [30]. The internal consistency of the scale in Study 1 was $\alpha = 0.913$, and in Study 2 it was $\alpha = 0.912$.

Family socioeconomic status was assessed using the Family Affluence Scale (FAS III), comprising items that reflect market forces, economic trends, and technological advances, as well as cultural, social, and geographical norms in consumption across Europe [37,38]. The FAS III was used as a continuous scale, consisting of six items about owning your private bedroom, family car, computer, dishwasher, bathrooms, and family vacations abroad. Higher scores reflect higher family affluence (the sum score could reach from 0 to 13 points). The internal consistency of the scale in Study 1 was $\alpha = 0.449$, and in Study 2 it was $\alpha = 0.519$.

2.3. Ethical Considerations

Both studies were approved by the Bioethics Centre of the Lithuanian University of Health Sciences (Study 1 No. BEC-M-03, 24 November 2021; Study 2 No. BEC-M-05, 17 March 2022). Permission was sought from the authorities of each selected school to conduct the research within their premises. Two forms of informed consent were implemented in the study. First, passive written informed consent was utilized, where parents were informed about their children's participation through the school administration. The school administration was instructed to inform parents about the upcoming survey through electronic diaries or other communication methods preferred by the school. Parents had to notify the administration of their refusal to participate. Second, active verbal consent was obtained from the adolescents before the survey commenced. At the beginning of the study, the researchers presented the purpose of the study, the principles of confidentiality and anonymity, and the option to refuse participation. The anonymity of the research participants and the confidentiality of the data were ensured. The data were collected by trained researchers.

2.4. Data Analysis

Data were processed using MS Excel 2016 and analyzed using IBM SPSS Statistics for Windows, Version 28.0 [39]. A statistical significance level of $\alpha = 0.05$ ($p < 0.05$) was used. Cronbach's alpha coefficient was calculated to assess the internal consistency of the scales.

Pearson's chi-squared test was used to compare the prevalence of anxiety by gender, class, and other dichotomized psychological factors between the 2021 and 2022 cohorts. Student's *t*-test was used to compare family socioeconomic status (FAS III) means between anxiety groups. The main analysis to identify prognostic factors of anxiety was conducted using univariate and multivariable binary logistic regressions. The strength of associations was expressed in odds ratios (ORs) with 95% confidence intervals (CIs), while the model fits were assessed using Nagelkerke R^2.

3. Results

Detailed characteristics of the study sample are presented in Table 1. In both cohorts, the gender distribution aimed for parity at around 50 percent. Among the grades, 9th grade participation was slightly higher, constituting approximately 28–29 percent, while 11th grade involvement was the lowest, at around 21 percent. The average Family Affluence Scale (FAS) score ranged between 7 and 8 points.

Table 1. Main characteristics of the study sample.

Characteristics		2021 (Study 1) % (*n*) or m (SD)	2022 (Study 2) % (*n*) or m (SD)
Gender			
	Boys	51.6 (237)	51.1 (3353)
	Girls	48.4 (222)	48.9 (3213)
Grade			
	5th	26.3 (120)	25.4 (1687)
	7th	23.9 (109)	23.9 (1584)
	9th	28.1 (128)	29.3 (1947)
	11th	21.7 (99)	21.4 (1419)
FAS III (0–13)		7.06 (2.24)	7.51 (2.32)

FAS III—Family Affluence Scale, showing the family's socioeconomic status. Valid percentages were calculated without missing responses.

The descriptive results in Table 2 show that the prevalence of anxiety among adolescents in 2021 and 2022 had no statistically significant difference ($p = 0.515$). Almost a quarter of adolescents (23–24%) in the 2021 and 2022 groups had elevated levels of anxiety on the GAD-7 scale, indicating a clinically significant risk of anxiety.

Table 2. Anxiety prevalence in the 2021 and 2022 cohorts and demographic comparisons.

		2021			2022			
		Moderate–Severe Anxiety % (*n*)	No–Mild Anxiety % (*n*)	*p* ^	Moderate–Severe Anxiety % (*n*)	No–Mild Anxiety % (*n*)	*p* ^	*p* *
General population		23.0 (100)	77.0 (335)		24.4 (1528)	75.6 (4741)		0.515
Gender								
	Boys	13.6 (30)	86.4 (191)	<0.001	15.1 (480)	84.9 (2696)	<0.001	
	Girls	32.9 (70)	67.1 (143)		33.7 (1022)	66.3 (2009)		
Grade								
	5th	18.9 (21)	81.1 (90)	0.163	20.0 (313)	80.0 (1254)	<0.001	
	7th	19.6 (19)	80.4 (78)		26.1 (390)	73.9 (1103)		
	9th	29.4 (37)	70.6 (89)		25.7 (476)	74.3 (1375)		
	11th	23.7 (23)	76.3 (74)		25.7 (349)	74.3 (1009)		
FAS III		6.67 ± 2.24	7.15 ± 2.22	0.066	7.47 ± 2.38	7.54 ± 2.29	0.293	

^—associations of anxiety rates and demographic characteristics within the 2021 and 2022 cohorts; *—comparison of anxiety rates between the 2021 and 2022 cohorts. FAS—Family Affluence Scale. Valid percentages were calculated without missing responses.

After comparing anxiety levels based on demographic and social characteristics, a significant gender disparity appeared, with girls expressing moderate or severe anxiety at about double the prevalence of boys (33–34% compared to 14–15%). Statistically significant gender differences were evident in both the 2021 and 2022 cohorts ($p < 0.001$). Furthermore, the results revealed systematic differences related to age, indicating a higher prevalence of elevated anxiety among older adolescents. Notably, a statistically significant increase in anxiety levels was observed in the 2022 cohort when comparing 5th and 11th grade students. Additionally, a higher risk of anxiety was related to lower socioeconomic status in 2021 and 2022, although no statistically significant differences were observed ($p = 0.066$ and $p = 0.293$, respectively).

Prior to developing prognostic anxiety models, it was important to examine potential variations in psychological factors among the examined cohorts (Table 3). This step aimed to comprehend any inherent differences or similarities between the groups, recognizing their potential influence on the manifestation and experience of anxiety within each cohort. Subsequently, the observed cohorts did not display statistically significant differences when evaluating other psychological factors ($p > 0.05$). This study showed that almost one-third of Lithuanian teenagers (28–31%) experienced heightened stress, and one-fifth (20–22%) reported feelings of loneliness. Furthermore, one in ten adolescents (10–12%) exhibited a risk for problematic social media use. Additionally, more than 40% of young people expressed a perceived lack of peer support, and approximately half of the surveyed adolescents (49–53%) assessed their self-efficacy as low in problem solving and decision implementation.

Table 3. Comparison of psychological factors between the 2021 and 2022 cohorts.

		2021	2022	*p*
		% (*n*)	% (*n*)	
PSMU				
	Problematic use	11.5 (48)	10.3 (624)	0.423
	Non-problematic use	88.5 (370)	89.7 (5463)	
Stress				
	High level	30.7 (138)	28.4 (1821)	0.292
	Low level	69.3 (311)	71.6 (4588)	
Loneliness				
	High	20.0 (92)	21.7 (1422)	0.394
	Low	80.0 (367)	78.3 (5119)	
Peer support				
	Low	44.5 (196)	42.9 (2746)	0.495
	High	55.5 (244)	57.1 (3658)	
Self-efficacy: problem solving				
	Low	50.8 (232)	48.7 (3186)	0.398
	High	49.2 (225)	51.3 (3353)	
Self-efficacy: decision implementation				
	Low	52.5 (238)	52.0 (3394)	0.824
	High	47.5 (215)	48.0 (3133)	

PSMU—problematic social media use. Valid percentages were calculated without missing responses.

Finally, to investigate the potential relationship between anxiety and selected psychological factors within the 2021 cohort, univariate regression analyses were conducted (Table 4). Substantial associations emerged between anxiety and specific factors. Notably, high stress levels (OR = 6.46, 95% CI 3.96–10.51), problematic social media use (OR = 3.75, 95% CI 1.97–7.13), and female gender (OR = 3.12, 95% CI 1.9–5.03). High loneliness levels (OR = 3.82, 95% CI 2.31–6.33), low peer support (OR = 1.71, 95% CI 1.08–2.69), and low self-efficacy in decision implementation (OR = 1.97, 95% CI 1.24–3.15) demonstrated significant associations with anxiety. However, in the multivariable model, these associations lost statistical significance. Self-efficacy (problem solving), grade, and socioeconomic status (FAS III) were not significantly associated with anxiety in the univariate model.

Table 4. Regression models predicting risk for moderate–severe anxiety in 2021.

		2021					
		Univariate Regression Models			Multivariable Regression Models Nagelkerke R^2 *		0.326
		OR	95% CI	p	OR	95% CI	p
Stress							
	Low level	1.00			1.00		
	High level	6.46	3.96–10.51	<0.001	5.89	3.11–11.17	<0.001
Loneliness							
	Low	1.00			1.00		
	High	3.82	2.31–6.33	<0.001	1.71	0.88–3.32	0.112
Social media use							
	Non-problematic	1.00			1.00		
	Problematic	3.75	1.97–7.13	<0.001	4.58	1.89–10.58	<0.001
Self-efficacy: problem solving							
	High	1.00			1.00		
	Low	1.56	0.99–2.45	0.055	0.62	0.33–1.18	0.146
Self-efficacy: decision implementation							
	High	1.00			1.00		
	Low	1.97	1.24–3.15	0.004	0.95	0.50–1.80	0.949
Peer support							
	High	1.00			1.00		
	Low	1.71	1.08–2.69	0.021	1.55	0.86–2.80	0.148
Gender							
	Boys	1.00			1.00		
	Girls	3.12	1.9–5.03	<0.001	2.87	1.58–5.22	0.001
Grade							
	5th	1.00					
	7th	1.04	0.52–2.08	0.903	1.18	0.49–2.88	0.710
	9th	1.78	0.97–3.28	0.064	2.16	0.98–4.77	0.057
	11th	1.33	0.68–2.60	0.399	1.38	0.58–3.30	0.466
FAS III		0.91	0.82–1.00	0.065	0.99	0.87–1.14	0.949

*—Nagelkerke R^2 of multivariable regression models. FAS III—Family Affluence Scale.

To investigate whether anxiety-related factors remain significant after controlling for competing indicators, two multivariable binary logistic regression models were created (Table 4) to identify independent variables that predict the likelihood of moderate–severe anxiety in 2021. The created model fit the data well: the likelihood ratio test $\chi^2 = 91.21$, $p < 0.001$; the Hosmer–Lemeshow test $\chi^2 = 9.76$, $p = 0.282$; and the Nagelkerke coefficient of pseudo determination was 0.326. The model correctly classified 92.8% without anxiety and 41.8% with moderate–severe anxiety. The overall percentage of correct predictions of the model was 80.2%.

The multivariable regression model predicting anxiety in the year 2021 revealed that high-level stress increased the probability of anxiety in adolescents by 5.89 times (95% CI 3.11–11.17). PSMU demonstrated a substantial effect on anxiety, with an OR of 4.58 (95% CI 1.89–10.58). In this model, girls were at higher risk of experiencing anxiety (OR = 2.87, 95% CI 1.58–5.22). All other indicators did not demonstrate significant associations with anxiety in the multivariable model. Overall, stress, problematic social media use, and gender emerged as significant predictors of anxiety in this cohort, with the strongest factor being stress.

Analyzing the data in the 2022 cohort, all included variables showed a significant association with anxiety in the univariate model (Table 5). The multivariable binary logistic regression model for predicting the likelihood of moderate–severe anxiety in 2022 fit the data well: the likelihood ratio test $\chi^2 = 1245.99$, $p < 0.001$; the Hosmer–Lemeshow test $\chi^2 = 11.29$, $p = 0.186$; and Nagelkerke coefficient of pseudo determination was 0.305. The model correctly classified 92.2% without anxiety and 41.0% with moderate–severe anxiety. The overall percentage of correct predictions of the model is 79.5%.

Table 5. Regression models predicting risk for moderate–severe anxiety in 2022.

		2022					
		Univariate Regression Models			Multivariable Regression Models Nagelkerke R^2 * 0.326		
		OR	**95% CI**	***p***	**OR**	**95% CI**	***p***
Stress							
	Low level	1.00			1.00		
	High level	7.06	6.21–8.03	<0.001	3.68	3.14–4.30	<0.001
Loneliness							
	Low	1.00			1.00		
	High	6.05	5.31–6.90	<0.001	2.85	2.43–3.35	<0.001
Social media use							
	Non-problematic	1.00			1.00		
	Problematic	2.96	2.48–3.53	<0.001	1.88	1.51–2.33	<0.001
Self-efficacy: problem solving							
	High	1.00			1.00		
	Low	2.24	1.98–2.52	<0.001	1.40	1.20–1.64	<0.001
Self-efficacy: decision implementation							
	High	1.00			1.00		
	Low	2.24	1.98–2.53	<0.001	1.13	0.96–1.32	0.137
Peer support							
	High	1.00			1.00		
	Low	1.37	1.22–1.54	<0.001	1.09	0.94–1.26	0.246
Gender							
	Boys	1.00			1.00		
	Girls	2.86	2.52–3.23	<0.001	2.26	1.94–2.62	<0.001
Grade							
	5th	1.00			1.00		
	7th	1.42	1.20–1.68	<0.001	1.26	1.02–1.57	0.033
	9th	1.39	1.18–1.63	<0.001	1.33	1.08–1.63	0.007
	11th	1.39	1.16–1.65	<0.001	1.43	1.15–1.79	0.002
FAS III		0.99	0.96–1.01	<0.001	1.07	1.04–1.11	<0.001

*—Nagelkerke R^2 of multivariable regression models. FAS III—Family Affluence Scale.

In the 2022 multivariable regression model, heightened stress increased the probability of experiencing anxiety by 3.68 times (95% CI 3.14–4.30), while problematic social media use increased the probability by 1.88 times (95% CI 1.51–2.33). Unlike in 2021, this model confirmed that loneliness significantly increased the probability of anxiety by 2.85 times (95% CI 2.43–3.35), and low self-efficacy (problem solving) also increased the probability of anxiety but was weaker (OR = 1.40; 95% CI 1.20–1.60). Additionally, older teenagers exhibited an increased risk of experiencing anxiety across different grades: 7th (OR = 1.26, 95% CI 1.02–1.57), 9th (OR = 1.33, 95% CI 1.08–1.63), and 11th (OR = 1.43, 95% CI 1.15–1.79). This trend was consistent with the higher likelihood observed among girls (OR = 2.26, 95% CI 1.94–2.62) and adolescents from lower socioeconomic status families (OR = 1.07, 95% CI 1.04–1.11).

Thus, multivariable models suggest that high levels of stress, problematic social media use, and female gender were consistent independent predictors of anxiety in the 2021 and 2022 cohorts. For these indicators, the effect sizes in terms of odds ratios were larger in 2021 compared to 2022. In contrast, loneliness and self-efficacy (problem solving) in 2022 were stronger predictors for anxiety compared to 2021.

4. Discussion

Anxiety is a common emotional state among children and adolescents during global crisis periods [4]. The COVID-19 pandemic and the Russian invasion of Ukraine have heightened global tensions, potentially leading to increased anxiety levels. Nevertheless, research on adolescent mental health during these difficult years, particularly in northeastern Europe, remains limited. Therefore, this study aimed to assess anxiety levels in Lithuanian adolescents during two key timeframes: October 2021 (marking the second COVID-19 wave) and April–June 2022 (coinciding with the onset of the Russian invasion of Ukraine). Furthermore, we have examined the potential influence of various psychological factors, including loneliness, problematic social media use, stress, and self-efficacy, on anxiety levels during these specific time periods.

Our research revealed that around 23–24% of adolescents in 2021 and 2022 showed elevated anxiety levels, indicating a clinically significant risk of anxiety. Notably, within the 2022 cohort, this increase was particularly notable among older students. Although there was no statistically significant difference in anxiety rates between 2021 and 2022, overall trends suggest higher rates compared to previous studies during calmer periods, where adolescent anxiety levels were typically around 6% [4]. Similar findings were noted in recent Finnish research examining trends in generalized anxiety among adolescents from 2013 to 2021 using the GAD-7 questionnaire. From 2013 to 2019, anxiety levels for males aged 13 to 17 years ranged between 5 and 6%, and for females in the same age group, it varied from 15 to 20% [40]. However, in 2021, anxiety levels increased to 7% for males and 30% for females. In our study, female participants showed a tendency to experience heightened anxiety more than twice as often as their male counterparts, both in 2021 and 2022. This difference remained statistically significant, even after adjustment for additional factors in the models (OR 2.87 and 2.26, respectively).

We found that around one-third of Lithuanian adolescents felt heightened stress levels, which may also lead to increased anxiety. Indeed, in our study, stress has demonstrated significant correlations with anxiety in univariate regressions and appeared as the most influential component in both the 2021 and 2022 multivariable regression models, with odds ratios of 5.9 and 3.7, respectively. This suggests that students experiencing higher stress levels were 4–6 times more likely to experience anxiety, independent from other included factors. This negative impact of high perceived stress levels on increasing anxiety, depression, and other mental health risk factors has been documented in prior research [41,42]. Feelings of a lack of control, a belief that one cannot effectively manage emerging challenges, and an excessive internal locus of control can trigger anxiety and fear of unpredictable events. In recent years, this connection between anxiety and stress among young people may have been more pronounced than before. The COVID-19 pandemic, as supported by other studies, increased stress and subsequently anxiety levels by disrupting the daily routines of young individuals, limiting social interactions and intensifying feelings of isolation, thereby raising concerns about virus transmission [43,44]. Furthermore, with the onset of the conflict in Ukraine, the stress levels of neighboring countries escalated even further. Concerns about the recent war in Ukraine were found to have increased in Polish [45], Romanian [27], and Czech youth [26]. The frequency of news consumption during times of military conflict played a significant role in exacerbating stress and anxiety [26,46,47]. Broadcasting the unpredictability of the conflict's outcome, potential threats to other countries, and witnessing death and violence could further lead to secondary trauma.

Hence, social media, recognized as a primary source for accessing news in the contemporary world, particularly among the youth, has been also identified as a potential contributor to heightened anxiety levels in this research. In our study cohorts, problematic social media use was evident in 11.5% (in 2021) and 10.3% (in 2022) of young individuals, with no significant difference between cohorts. PSMU has also emerged as a significant predictive factor for anxiety in our samples. According to logistic regression models, youths exhibiting problematic social media use were associated with 4.9 times higher odds of experiencing elevated anxiety in 2021 and 1.9 times higher odds in 2022. Prior research has shown that PSMU is linked to sleep disturbances and reduced life satisfaction [48], as well as depression, anxiety, and stress [49]. Various hypotheses can be formulated regarding the connection between PSMU and anxiety during these periods, including higher anxiety due to exposure to distressing news on social media [25], the 'fear of missing out' [50], or the possibility that the addictive behavior itself is associated with anxiety (the inability to stay away from social media, which is one of the symptoms of PSMU, may trigger anxiety). On the other hand, it is important to note that our study was cross-sectional, and the relationship between anxiety and PSMU may be bidirectional—individuals experiencing anxiety may be more prone to engage in PSMU [51].

Almost half of the surveyed adolescents reported low self-efficacy which, in turn, increased odds of experiencing anxiety. In our study, self-efficacy was examined by two items related to problem solving and decision implementation, but only the former exhibited statistical significance within the multivariable regression model in 2022. We found that adolescents who think that they usually cannot find a solution to an emerging challenge exhibit a 1.4-fold increased likelihood of experiencing heightened anxiety. Similar results demonstrating links between low self-efficacy and anxiety were observed in previous studies [52,53]. Interestingly, the question about decision implementation in our study demonstrated statistical significance in bivariate analysis but lost this significance when controlling for other variables. This suggests that the link between increased anxiety and self-efficacy is more related to how individuals perceive their ability to come up with solutions rather than whether they put those solutions into practice. Hence, the affective dimension of decision-making assumes greater salience. It is worth noting that such an association did not manifest within the 2021 cohort.

Our research findings indicated that 20–22% of the surveyed students reported feeling lonely most of the time. Over the past decade, there has been a rapid increase in loneliness among young people [54]. This trend could have been accelerated further by the onset of the global COVID-19 pandemic, which imposed social restrictions and may have increased the number of people feeling lonely [55]. In our study, loneliness was also significantly linked to increased levels of anxiety. In 2022, students who reported feeling lonely were nearly three times more likely to experience heightened anxiety, even after controlling for other factors. A similar trend was noted in the 2021 cohort, although the associations did not reach statistical significance. These results were in line with the recent literature review revealing that higher levels of loneliness during the COVID-19 pandemic were significantly associated with poorer well-being, including higher depression symptoms, anxiety symptoms, gaming addiction, and sleep problems [56].

Peer support was another factor that was analyzed in our study. We found that low peer support was present in slightly more than 42% of both study cohorts, and there was no statistically significant difference between years. Low social support showed significant correlations with higher anxiety levels in univariate regressions; however, when other factors were included in the models, these associations became non-significant. Research indicates that peer support is one of the most significant factors, after parental support, which can influence a child's psychological well-being [57]. Peer conflicts, victimization, or isolation can increase suicidal ideation and anxiety [58]. One possible explanation for the lack of the significance of peer support in multivariable regression models is the inclusion of loneliness as a variable, which may reduce the role of social support in predicting anxiety. Indeed, studies suggest that loneliness serves as a mediator between social support and life satisfaction: when a person has social support, they feel less lonely, and as a result, their life satisfaction increases [59].

Upon summarizing both studies, it became evident that anxiety levels remained comparable between 2021 and 2022, aligning with the respective prognostic factors. This similarity may be attributed to the assessment of generalized anxiety disorder symptoms using the GAD-7 questionnaire rather than focusing solely on event-specific anxiety. Such an observation suggests that a general anxiety assessment could yield a similar predictive model across both periods. However, the multivariate regressions revealed variations in the impact degrees of prognostic factors. In 2021, high stress, social media use, and female gender emerged as more prominent predictors. Conversely, in 2022, loneliness and problem solving self-efficacy became stronger predictors for anxiety compared to 2021. Several hypotheses could be derived from these findings. Plausibly, COVID-19 might have exerted a more substantial influence on anxiety levels than the impact of war during the respective periods. Additionally, the year 2022 possibly witnessed a heightened adaptation to the persistent threat of COVID-19. Factors contributing to this adaptation include the prolonged duration since the pandemic's onset, advancements in disease control through vaccination and treatment, the facilitation of open school settings, and a shift in public focus toward the war. These adjustments potentially influenced the behavioral dynamics of predictive factors associated with anxiety, thereby contributing to the observed differences in odds ratios. Nevertheless, a comprehensive exploration of these hypotheses necessitates longitudinal study designs. Such designs would enable a more detailed investigation into the causal relationships and temporal dynamics between predictive factors and anxiety levels across varying timeframes.

There are several limitations of our research that need to be addressed. First, the cross-sectional nature of this study restricts the ability to establish causal relationships between analyzed variables. Therefore, it is important to emphasize that regression models, while predicting anxiety, may also exhibit some reverse or bidirectional associations between dependent and independent variables. The main aim of our study was to explore the anxiety rates and their correlates among Lithuanian adolescents during two distinct time periods: the second wave of the COVID-19 pandemic and the onset of the Russian invasion of Ukraine. However, it is essential to acknowledge that both studies were separated by half a year, which implies that residual COVID-related influences on overall anxiety levels and other indicators may have persisted within the 2022 cohort as well. On the other hand, it is worth noting that the potential lasting effects of the COVID-19 pandemic may remain relevant in the years to come, making the elimination of this background influence a challenging task. It is also noteworthy that the examined cohorts had considerably different sample sizes. While statistically significant associations between the psychological characteristics of cohorts were not found, the statistical significance in the regression models of the 2021 cohort could likely have been the result of a smaller sample size. Finally, our study exclusively incorporated self-report questionnaires. More objective assessments or data from diverse sources, such as parents, teachers, and healthcare professionals, were not collected. Therefore, the findings regarding anxiety expression should be approached with caution. On the other hand, previous research has demonstrated the validity and sensitivity of the GAD-7 questionnaire in assessing anxiety risk [32]. Additionally, recent work by Stromájer and colleagues [60] has highlighted discrepancies between objective stress measurements, such as cortisol levels, and subjective evaluations. This discrepancy emphasizes that while objective measures might not always align with individuals' subjective experiences, subjective tests used to measure anxiety might offer a more accurate indication of anxiety levels and reflect the inner world of adolescents.

5. Conclusions

Anxiety represents a common response among adolescents to life's unpredictability, which may be particularly exacerbated during periods of global upheaval. Our study demonstrated that approximately 23–24% of Lithuanian adolescents experienced significantly elevated anxiety levels during the second wave of the COVID-19 pandemic and the Ukraine conflict. Moreover, we found that the risk of elevated anxiety levels increased overall stress, decreased self-efficacy, loneliness, or problematic social media use. Additionally, we identified that certain demographic characteristics, including female gender, older age, and lower socioeconomic status, contributed to heightened anxiety risk.

This study not only contributes to the theoretical understanding of anxiety but also offers practical insights. Conducted within a school environment, our research unveiled a significant correlation between anxiety-provoking events and their profound impact on the mental well-being of young individuals. This impact often manifests as heightened levels of anxiety, increased stress, and a noticeable dependence on social media. Therefore, this provides a foundation for educational and health institutions to develop prevention and intervention programs supporting adolescent psychological health during periods of global crises or challenging circumstances. In our research, specific vulnerable groups to anxiety emerged, notably girls, older students, and those from lower socioeconomic backgrounds. Targeted attention toward these cohorts is essential within school settings to identify mental health concerns and provide tailored support. Additionally, our study illuminates the complex relationship between anxiety and various other psychological indicators, emphasizing the importance of comprehensive assessments and integrated approaches in designing and implementing psychological interventions.

Author Contributions: L.Š.-N. and K.Š. developed the research concept. L.Š.-N. conducted data analysis, drafted the introduction, and interpreted the results in the discussion section. M.Ž. contributed to the methodology and the results sections. K.Š. supervised manuscript preparation. Both L.Š.-N. and K.Š. critically revised and edited the final manuscript. All authors have read and agreed to the published version of the manuscript.

Funding: This research was funded by the World Health Organization's office in Lithuania.

Institutional Review Board Statement: Both studies involved in this paper were conducted in accordance with the Declaration of Helsinki and approved by the Bioethics Centre of the Lithuanian University of Health Sciences (Study 1 No. BEC-M-03, 24 November 2021; Study 2 No. BEC-M-05, 17 March 2022).

Informed Consent Statement: All subjects involved in this study provided verbal informed consent, while a passive informed consent procedure was employed to inform participants' parents about the study and provide them with the opportunity to refuse participation.

Data Availability Statement: The data presented in this study are available upon request from the corresponding author. The data are not publicly available due to their containing information that could compromise the privacy of research participants.

Acknowledgments: The authors gratefully acknowledge the HBSC Lithuanian team for collecting the data used in this study.

Conflicts of Interest: The authors declare no conflicts of interest.

References

1. Miller, K.E.; Rasmussen, A. War exposure, daily stressors, and mental health in conflict and post-conflict settings: Bridging the divide between trauma-focused and psychosocial frameworks. *Soc. Sci. Med.* **2010**, *70*, 7–16. [CrossRef]
2. Pierce, M.; Hope, H.; Ford, T.; Hatch, S.; Hotopf, M.; John, A.; Kontopantelis, E.; Webb, R.; Wessely, S.; McManus, S.; et al. Mental health before and during the COVID-19 pandemic: A longitudinal probability sample survey of the UK population. *Lancet Psychiatry* **2020**, *7*, 883–892. [CrossRef]
3. Magson, N.R.; Freeman, J.Y.A.; Rapee, R.M.; Richardson, C.E.; Oar, E.L.; Fardouly, J. Risk and Protective Factors for Prospective Changes in Adolescent Mental Health during the COVID-19 Pandemic. *J. Youth Adolesc.* **2021**, *50*, 44–57. [CrossRef]
4. Polanczyk, G.V.; Salum, G.A.; Sugaya, L.S.; Caye, A.; Rohde, L.A. Annual Research Review: A meta-analysis of the worldwide prevalence of mental disorders in children and adolescents. *J. Child Psychol. Psychiatry* **2015**, *56*, 345–365. [CrossRef]
5. Merikangas, K.R.; He, J.-P.; Burstein, M.; Swanson, S.A.; Avenevoli, S.; Cui, L.; Benjet, C.; Georgiades, K.; Swendsen, J. Lifetime Prevalence of Mental Disorders in U.S. Adolescents: Results from the National Comorbidity Survey Replication–Adolescent Supplement (NCS-A). *J. Am. Acad. Child Adolesc. Psychiatry* **2010**, *49*, 980–989. [CrossRef]
6. Erskine, H.E.; Moffitt, T.E.; Copeland, W.E.; Costello, E.J.; Ferrari, A.J.; Patton, G.; Degenhardt, L.; Vos, T.; Whiteford, H.A.; Scott, J.G. A heavy burden on young minds: The global burden of mental and substance use disorders in children and youth. *Psychol. Med.* **2014**, *45*, 1551–1563. [CrossRef]
7. de Lijster, J.M.; Dieleman, G.C.; Utens, E.M.; Dierckx, B.; Wierenga, M.; Verhulst, F.C.; Legerstee, J.S. Social and academic functioning in adolescents with anxiety disorders: A systematic review. *J. Affect. Disord.* **2018**, *230*, 108–117. [CrossRef]
8. Raknes, S.; Pallesen, S.; Himle, J.A.; Bjaastad, J.F.; Wergeland, G.J.; Hoffart, A.; Dyregrov, K.; Håland, Å.T.; Haugland, B.S.M. Quality of life in anxious adolescents. *Child Adolesc. Psychiatry Ment. Health* **2017**, *11*, 33. [CrossRef]
9. Essau, C.A.; Lewinsohn, P.M.; Olaya, B.; Seeley, J.R. Anxiety disorders in adolescents and psychosocial outcomes at age 30. *J. Affect. Disord.* **2014**, *163*, 125–132. [CrossRef]
10. Liu, Y.; Yue, S.; Hu, X.; Zhu, J.; Wu, Z.; Wang, J.; Wu, Y. Associations between feelings/behaviors during COVID-19 pandemic lockdown and depression/anxiety after lockdown in a sample of Chinese children and adolescents. *J. Affect. Disord.* **2021**, *284*, 98–103. [CrossRef]

11. Orgilés, M.; Espada, J.P.; Delvecchio, E.; Francisco, R.; Mazzeschi, C.; Pedro, M.; Morales, A. Anxiety and Depressive Symptoms in Children and Adolescents during COVID-19 Pandemic: A Transcultural Approach. *Psicothema* **2021**, *33*, 125–130. [CrossRef]
12. Śniadach, J.; Szymkowiak, S.; Osip, P.; Waszkiewicz, N. Increased Depression and Anxiety Disorders during the COVID-19 Pandemic in Children and Adolescents: A Literature Review. *Life* **2021**, *11*, 1188. [CrossRef]
13. Marques de Miranda, D.; da Silva Athanasio, B.; Sena Oliveira, A.C.; Simoes-e-Silva, A.C. How is COVID-19 pandemic impacting mental health of children and adolescents? *Int. J. Disaster Risk Reduct.* **2020**, *51*, 101845. [CrossRef]
14. Ma, L.; Mazidi, M.; Li, K.; Li, Y.; Chen, S.; Kirwan, R.; Zhou, H.; Yan, N.; Rahman, A.; Wang, W.; et al. Prevalence of mental health problems among children and adolescents during the COVID-19 pandemic: A systematic review and meta-analysis. *J. Affect. Disord.* **2021**, *293*, 78–89. [CrossRef]
15. Rodman, A.M.; Rosen, M.L.; Kasparek, S.W.; Mayes, M.; Lengua, L.; Meltzoff, A.N.; McLaughlin, K.A. Social behavior and youth psychopathology during the COVID-19 pandemic: A longitudinal study. *Dev. Psychopathol.* **2022**, 1–13. [CrossRef]
16. Rossi, L.; Behme, N.; Breuer, C. Physical Activity of Children and Adolescents during the COVID-19 Pandemic—A Scoping Review. *Int. J. Environ. Res. Public Health* **2021**, *18*, 11440. [CrossRef]
17. Gruhn, M.; Miller, A.B.; Machlin, L.; Motton, S.; Thinzar, C.E.; Sheridan, M.A. Child Anxiety and Depression Symptom Trajectories and Predictors over 15 Months of the Coronavirus Pandemic. *Res. Child Adolesc. Psychopathol.* **2023**, *51*, 233–246. [CrossRef]
18. Cauberghe, V.; Van Wesenbeeck, I.; De Jans, S.; Hudders, L.; Ponnet, K. How Adolescents Use Social Media to Cope with Feelings of Loneliness and Anxiety During COVID-19 Lockdown. *Cyberpsychol. Behav. Soc. Netw.* **2021**, *24*, 250–257. [CrossRef]
19. Marciano, L.; Ostroumova, M.; Schulz, P.J.; Camerini, A.-L. Digital Media Use and Adolescents' Mental Health During the COVID-19 Pandemic: A Systematic Review and Meta-Analysis. *Front. Public Health* **2022**, *9*, 793868. [CrossRef]
20. Boursier, V.; Gioia, F.; Musetti, A.; Schimmenti, A. Facing Loneliness and Anxiety during the COVID-19 Isolation: The Role of Excessive Social Media Use in a Sample of Italian Adults. *Front. Psychiatry* **2020**, *11*, 586222. [CrossRef] [PubMed]
21. Pfefferbaum, B.; North, C.S. Mental Health and the COVID-19 Pandemic. *N. Engl. J. Med.* **2020**, *383*, 510–512. [CrossRef]
22. Danzer, A.M.; Danzer, N. The long-run consequences of Chernobyl: Evidence on subjective well-being, mental health and welfare. *J. Public Econ.* **2016**, *135*, 47–60. [CrossRef]
23. Havenaar, J.M.; Van den Brink, W.; Van den Bout, J.; Kasyanenko, A.P.; Poelijoe, N.W.; Wholfarth, T.; Meijler-Iljina, L.I. Mental health problems in the Gomel region (Belarus): An analysis of risk factors in an area affected by the Chernobyl disaster. *Psychol. Med.* **1996**, *26*, 845–855. [CrossRef]
24. Cwikel, J.G.; Abdelgani, A.; Rozovski, U.; Kordysh, E.; Goldsmith, J.R.; Quastel, M.R. Long-Term stress reactions in new immigrants to israel exposed to the chernobyl accident. *Anxiety Stress. Coping* **2000**, *13*, 413–439. [CrossRef]
25. Riad, A.; Drobov, A.; Krobot, M.; Antalová, N.; Alkasaby, M.A.; Peřina, A.; Koščík, M. Mental Health Burden of the Russian–Ukrainian War 2022 (RUW-22): Anxiety and Depression Levels among Young Adults in Central Europe. *Int. J. Environ. Res. Public Health* **2022**, *19*, 8418. [CrossRef]
26. Riad, A.; Drobov, A.; Alkasaby, M.A.; Peřina, A.; Koščík, M. Nuclear Anxiety Amid the Russian-Ukrainian War 2022 (RUW-22): Descriptive Cross-Sectional Study. *Int. J. Environ. Res. Public Health* **2023**, *20*, 3551. [CrossRef]
27. Maftei, A.; Dănilă, O.; Măirean, C. The war next-door—A pilot study on Romanian adolescents' psychological reactions to potentially traumatic experiences generated by the Russian invasion of Ukraine. *Front. Psychol.* **2022**, *13*, 1051152. [CrossRef]
28. Pfefferbaum, B.; Newman, E.; Nelson, S.D.; Nitiéma, P.; Pfefferbaum, R.L.; Rahman, A. Disaster Media Coverage and Psychological Outcomes: Descriptive Findings in the Extant Research. *Curr. Psychiatry Rep.* **2014**, *16*, 464. [CrossRef] [PubMed]
29. Runze, J.; Marten, F.; Brinke, L.T. The effect of exposure to social-media coverage of the Russo-Ukrainian war on stress symptoms in Dutch adolescents. *PsyArXiv* **2022**. [CrossRef]
30. Inchley, J.; Currie, D.; Piper, A.; Jåstad, A.; Cosma, A.; Nic Gabhainn, S.; Samdal, O. *Health Behaviour in School-Aged Children (HBSC) Study Protocol: Background, Methodology, Mandatory Questions and Optional Packages for the 2021/22 Survey*; MRC/CSO Social and Public Health Sciences Unit, University of Glasgow: Glasgow, UK, 2021.
31. Spitzer, R.L.; Kroenke, K.; Williams, J.B.W.; Löwe, B. A Brief Measure for Assessing Generalized Anxiety Disorder: The GAD-7. *Arch. Intern. Med.* **2006**, *166*, 1092–1097. [CrossRef] [PubMed]
32. Mossman, S.A.; Luft, M.J.; Schroeder, H.K.; Varney, S.T.; Fleck, D.E.; Barzman, D.H.; Gilman, R.; DelBello, M.P.; Strawn, J.R. The Generalized Anxiety Disorder 7-item scale in adolescents with generalized anxiety disorder: Signal detection and validation. *Ann. Clin. Psychiatry Off. J. Am. Acad. Clin. Psychiatr.* **2017**, *29*, 227–234A.
33. van den Eijnden, R.J.J.M.; Lemmens, J.S.; Valkenburg, P.M. The Social Media Disorder Scale. *Comput. Hum. Behav.* **2016**, *61*, 478–487. [CrossRef]
34. Cohen, S.; Kamarck, T.; Mermelstein, R. A global measure of perceived stress. *J. Health Soc. Behav.* **1983**, *24*, 385–396. [CrossRef] [PubMed]
35. Vancampfort, D.; Ashdown-Franks, G.; Smith, L.; Firth, J.; Van Damme, T.; Christiaansen, L.; Stubbs, B.; Koyanagi, A. Leisure-time sedentary behavior and loneliness among 148,045 adolescents aged 12–15 years from 52 low- and middle-income countries. *J. Affect. Disord.* **2019**, *251*, 149–155. [CrossRef] [PubMed]
36. Zimet, G.D.; Dahlem, N.W.; Zimet, S.G.; Farley, G.K. The Multidimensional Scale of Perceived Social Support. *J. Persinal. Assess.* **1988**, *52*, 30–41. [CrossRef]
37. Hartley, J.E.K.; Torsheim, T.; Currie, C.; Baban, A.; Cavallo, F.; Dankulincová, Z.; Due, P.; Gajewski, J.; Geckova, A.; Holstein, B.; et al. *FAS III Development Study Report*; St. Andrews University: St. Andrews, UK, 2013.

38. Torsheim, T.; Cavallo, F.; Levin, K.A.; Schnohr, C.; Mazur, J.; Niclasen, B.; Currie, C.; FAS Development Study Group. Psychometric validation of the Revised Family Affluence Scale: A latent variable approach. *Child Indic. Res.* **2016**, *9*, 771–784. [CrossRef]
39. IBM Corp. *IBM SPSS Statistics for Windows, Version 28.0*; IBM Corp.: Armonk, NY, USA, 2021.
40. Kiviruusu, O.; Haravuori, H.; Lindgren, M.; Therman, S.; Marttunen, M.; Suvisaari, J.; Aalto-Setälä, T. Generalized anxiety among Finnish youth from 2013 to 2021—Trend and the impact of COVID-19. *J. Affect. Disord.* **2023**, *330*, 267–274. [CrossRef]
41. Beaver, K.M.; Vaughn, M.G.; Wright, J.P.; DeLisi, M. An interaction between perceived stress and 5HTTLPR genotype in the prediction of stable depressive symptomatology. *Am. J. Orthopsychiatry* **2012**, *82*, 260–266. [CrossRef]
42. Liu, Q.; Wang, Z. Perceived stress of the COVID-19 pandemic and adolescents' depression symptoms: The moderating role of character strengths. *Pers. Individ. Differ.* **2021**, *182*, 111062. [CrossRef]
43. Mohler-Kuo, M.; Dzemaili, S.; Foster, S.; Werlen, L.; Walitza, S. Stress and Mental Health among Children/Adolescents, Their Parents, and Young Adults during the First COVID-19 Lockdown in Switzerland. *Int. J. Environ. Res. Public Health* **2021**, *18*, 4668. [CrossRef]
44. Hoyt, L.T.; Cohen, A.K.; Dull, B.; Castro, E.M.; Yazdani, N. "Constant Stress Has Become the New Normal": Stress and Anxiety Inequalities Among U.S. College Students in the Time of COVID-19. *J. Adolesc. Health* **2020**, *68*, 270–276. [CrossRef]
45. Wójtowicz-Szefler, M.; Grzankowska, I.; Deja, M. The mental condition of Polish adolescents during the COVID-19 pandemic and war in Ukraine. *Front. Public Health* **2023**, *11*, 1257384. [CrossRef]
46. Malka, V.; Ariel, Y.; Avidar, R. Fighting, worrying and sharing: Operation 'Protective Edge' as the first WhatsApp war. *Media War Confl.* **2015**, *8*, 329–344. [CrossRef]
47. Bodas, M.; Siman-Tov, M.; Peleg, K.; Solomon, Z. Anxiety-Inducing Media: The Effect of Constant News Broadcasting on the Well-Being of Israeli Television Viewers. *Psychiatry* **2015**, *78*, 265–276. [CrossRef]
48. Buda, G.; Lukoševičiūtė, J.; Šalčiūnaitė, L.; Šmigelskas, K. Possible Effects of Social Media Use on Adolescent Health Behaviors and Perceptions. *Psychol. Rep.* **2021**, *124*, 1031–1048. [CrossRef]
49. Shannon, H.; Bush, K.; Villeneuve, P.J.; Hellemans, K.G.; Guimond, S. Problematic Social Media Use in Adolescents and Young Adults: Systematic Review and Meta-analysis. *JMIR Ment. Health* **2022**, *9*, e33450. [CrossRef]
50. Swan, A.J.; Kendall, P.C. Fear and missing out: Youth anxiety and functional outcomes. *Clin. Psychol. Sci. Pract.* **2016**, *23*, 417–435. [CrossRef]
51. Billieux, J.; Maurage, P.; Lopez-Fernandez, O.; Kuss, D.J.; Griffiths, M.D. Can Disordered Mobile Phone Use Be Considered a Behavioral Addiction? An Update on Current Evidence and a Comprehensive Model for Future Research. *Curr. Addict. Rep.* **2015**, *2*, 156–162. [CrossRef]
52. Muris, P. Relationships between self-efficacy and symptoms of anxiety disorders and depression in a normal adolescent sample. *Personal Individ. Differ.* **2002**, *32*, 337–348. [CrossRef]
53. Tahmassian, K.; Jalali Moghadam, N. Relationship between self-efficacy and symptoms of anxiety, depression, worry and social avoidance in a normal sample of students. *Child Indic. Res.* **2011**, *5*, 91–98.
54. Twenge, J.M.; Haidt, J.; Blake, A.B.; McAllister, C.; Lemon, H.; Le Roy, A. Worldwide increases in adolescent loneliness. *J. Adolesc.* **2021**, *93*, 257–269. [CrossRef]
55. Ernst, M.; Niederer, D.; Werner, A.M.; Czaja, S.J.; Mikton, C.; Ong, A.D.; Rosen, T.; Brähler, E.; Beutel, M.E. Loneliness before and during the COVID-19 pandemic: A systematic review with meta-analysis. *Am. Psychol.* **2022**, *77*, 660–677. [CrossRef]
56. Farrell, A.H.; Vitoroulis, I.; Eriksson, M.; Vaillancourt, T. Loneliness and Well-Being in Children and Adolescents during the COVID-19 Pandemic: A Systematic Review. *Children* **2023**, *10*, 279. [CrossRef]
57. Wu, Y.-J.; Lee, J. The most salient global predictors of adolescents' subjective Well-Being: Parental support, peer support, and anxiety. *Child Indic. Res.* **2022**, *15*, 1601–1629. [CrossRef]
58. Biswas, T.; Scott, J.G.; Munir, K.; Renzaho, A.M.N.; Rawal, L.B.; Baxter, J.; Mamun, A.A. Global variation in the prevalence of suicidal ideation, anxiety and their correlates among adolescents: A population based study of 82 countries. *EClinicalMedicine* **2020**, *24*, 100395. [CrossRef]
59. Kong, F.; You, X. Loneliness and Self-Esteem as Mediators Between Social Support and Life Satisfaction in Late Adolescence. *Soc. Indic. Res.* **2011**, *110*, 271–279. [CrossRef]
60. Stromájer, G.P.; Csima, M.; Iváncsik, R.; Varga, B.; Takács, K.; Stromájer-Rácz, T. Stress and Anxiety among High School Adolescents: Correlations between Physiological and Psychological Indicators in a Longitudinal Follow-Up Study. *Children* **2023**, *10*, 1548. [CrossRef]

MDPI AG
Grosspeteranlage 5
4052 Basel
Switzerland
Tel.: +41 61 683 77 34

Children Editorial Office
E-mail: children@mdpi.com
www.mdpi.com/journal/children

www.ingramcontent.com/pod-product-compliance
Lightning Source LLC
Chambersburg PA
CBHW042010260926
51850CB00020B/2969
* 9 7 8 3 7 2 5 8 3 6 3 5 2 *